Advances in Anatomy
Embryology and Cell Biology

Vol. 116

W0105901

J.E. Pauly, Little Rock T.H. Schiebler, Würzburg

Humio Mizoguti

A Fifteen-somite Human Embryo

With 57 Figures

Springer-Verlag Berlin Heidelberg GmbH

Prof. Humio Mizoguti, MD, PhD
Department of Anatomy
Kobe University School of Medicine
Kusunokicho 7-5-1, 650 Chuoku, Kobe, Japan

Library of Congress Cataloging-in-Publication Data
Mizoguti, Humio, 1925–
    A fifteen-somite human embryo/Humio Mizoguti.
    p. cm. —(Advances in anatomy, embryology, and cell biology; vol. 116)
    Bibliography: p.
    ISBN 978-3-540-50565-5   ISBN 978-3-642-74294-1 (eBook)
    DOI 10.1007/978-3-642-74294-1

    1. Embryology, Human. I. Title. II. Series: Advances in anatomy, embryology, and
    cell biology; v. 116.
    [DNLM: 1. Embryo. W1 AD433K v. 116/QS 604 M6852f]
    QL801.E67 vol. 116 [QM601] 574.4 s—dc19 [611'.013] DNLM/DLC 89-6071

The use of general descriptive names, trade names, trade marks, etc. in this
publication, even if the former are not especially identified, is not to be taken as a sign
that such names, as understood by the Trade Marks and Merchandise Marks Act,
may accordingly be used freely by anyone.

Product Liability: The publisher can give no guarantee for information about drug
dosage and application thereof contained in this book. In every individual case the
respective user must check its accuracy by consulting other pharmaceutical literature.

2121/3140-543210 – Printed on acid-free paper

Contents

1 Introduction

There have been numerous reports about human embryos in early developmental stages, among which those of Streeter and co-workers published in *Contributions to Embryology*, Carnegie Institution of Washington, have clarified the issue greatly. Recently O'Rahilly and Müller made a comprehensive survey of human embryos younger than 8 weeks, mainly based on the Carnegie collections.

However, because it is seldom possible to obtain a well-preserved embryo, detailed histological descriptions of embryos in Streeter's developmental horizon XI (Carnegie stage 11), having 13–20 paired somites, are rare, to the extent that every additional report on the embryo in this stage is welcome.

The aim of this paper is to provide a complete set of photomicrographs showing in detail the histological features of a very well preserved human embryo having 15 paired somites.

2 Specimen and Method

This embryo was found during a forensic dissection of a 38-year-old Japanese woman on December 4, 1982 in the Department of Forensic Medicine, Kobe University School of Medicine. The uterus was slightly enlarged and a marked corpus luteum was present in the left ovary. When the anterior wall of the uterus was opened, a gestational sac (GS) about 10 mm in diameter was seen. With the naked eye, however, one was unable to perceive anything in it. The GS was fixed in 10% formalin for 5 days and then examined in formalin with a binocular dissecting microscope. During removal of adnexa, the embryo was found intact, surrounded by a perfect amnion.

Before and after opening the amnion numerous photomicrographs were taken from various angles in black-and-white as well as in color.

The embryo was then washed with buffer, postfixed with 1% osmium tetroxide, dehydrated with graded alcohols, and embedded in an epoxy resin mixture. Serial transverse sections about 0.75 μm thick were made and stained with toluidine blue. At sectioning, the sectioning plane was adjusted as exactly as possible at right angles to the longitudinal axis of the embryo at every exchange of glass knives.

Photomicrographs of sections were taken with a BH-2 microscope (Olympus Optical Co., Tokyo, Japan) in combination with an S-Plan-Apochromat 20 (NA = 0.70) and photoprojection lens, NFK 2.5 LD. For purposes of adequate contrast, a narrow band interference filter (AL, 546 nm, Zeiss, Oberkochen, FRG), was used. At printing, negatives were enlarged three times with a Valoy II enlarger (Leitz, Wetzlar, FRG). The printed positives were put together to make composite photomicrographs. The total magnification on the final photomicrographs was 150 times.

3 Description of Specimen

3.1 General Form of Embryo (Figs. 1–7)

This embryo had, at greatest length, 4.1 mm measured in formalin through the amnion. The amniotic sac was relatively small, and the ventral half of the embryo directly faced the extra-embryonic coelom.

Viewed dorsally, 14 pairs of somites were discerned, and the neural tube appeared closed, approximately in the somite region (Fig. 7). The anterior and posterior neuropores were both still open into the amniotic cavity (Figs. 5, 7). The longitudinal axis of the embryo was roughly straight, but bent slightly to the right at the level of the tenth somite and twisted gently to the right at the most caudal portion of the embryo. This twist corresponded to the course of the body stalk which arose from the ventral aspect of this portion, ran first left-ventrally and soon turned left-dorsally (Figs. 2, 7).

Laterally observed, the embryo showed a slight but distinct dorsal flexure centered over the seventh and eighth somites (Fig. 5). The anterior half of the embryonic body anterior to this flexure was elevated by the prominent pericardiac swelling, beyond which the most anterior (cephalic) portion of the body bent sharply ventralwards, casuing a distinct hollow, the stomatodeum, between these two structures (Figs. 1–4, 6). The posterior half of the body was at first slightly elevated, then made a gentle ventral flexure centered over the 14th somite and ended with a small rounded tip (Figs. 1, 5, 7).

The ventral portion of the embryo consisted of a marked yolk sac, about 2.5 mm in diameter, whose surface was covered by a meshwork of highly developed blood vessels (Figs. 1, 2, 5, 6). Between the yolk sac and pericardiac swelling, in which the heart tube was perceived as a tortuous tubule, the septum transversum was conspicuously noted (Figs. 3–6).

In the most cephalic region of the embryo the neural tube was open into the amniotic cavity, and the right and left lips of the neural plate were widely separated from each other. The most cephalic portion of the neural plate evaginated laterally to form a marked optic evagination on each side, which overhung anteriorly to the stomatodeum, floored by the buccopharyngeal membrane having three perforations (Fig. 4). Posterior to the stomatodeum two branchial arches (the first prominent, the second less so),

3

and two branchial grooves were seen (Figs. 3, 5–7). The third branchial arch posterior to the second groove was faintly suggested. At the dorsal end of the second branchial groove a small round depression, the otic placode, was clearly recognizable (Figs. 3, 5–6). In an extent between the levels of the 5th–14th somite the intra-embryonic coelom was widely open laterally into the extra-embryonic one (Figs. 3, 6).

3.2 Serial Sections

The observation of serial sections first revealed that cells and tissues were very well preserved. Caudal to the 14th somite one more somite, the 15th, was recognized. Further posteriorly the mesoderm was no longer segmented.

3.2.1 Ectoderm

3.2.1.1 Central Nervous System

Closure of the neural tube took place from the level of the first branchial groove, corresponding to the level of the anterior one-third of the rhombencephalon where the facio-acoustic neural crest cell aggregation was observed (Fig. 18), to the level of about 0.4 mm posterior to the 15th somite, coming across the anterior end of the cloacal membrane (Fig. 50). The anterior neuropore thus opened at the level of the anterior one-third of the rhombencephalon.

The neural plate and the wall of the neural tube consisted exclusively of a tall pseudostratified columnar epithelium with 5–12 rows of nuclei, i.e., matrix, and the mantle and marginal layers did not differentiate anywhere throughout the central nervous system.

In the region of the midbrain the longitudinal axis of the neural plate bent ventrally to make the cephalic flexure so that the neural plate of the forebrain occupied a vertical position (Fig. 3). The most anterior (cephalic) portion of the neural plate, i.e., the ventral two-thirds of the forebrain, evaginated laterally to form a marked optic evagination on each side, whose lateral and ventral aspects were in contact with the overlying ectoderm which, however, showed no sign of differentiation of the lens placode (Figs. 8–10). The boundary between the fore- and midbrain was not identified, whereas that between the mid- and hindbrain (rhombencephalon) was faintly indicated by a shallow depression.

Between the levels of the first branchial groove and the first somite the rhombencephalon was closed dorsally by a thin two-cell-layer membrane, the tegmen rhombencephali (Figs. 17–26), which displayed its greatest width of about 130 μm just posterior to the otic pit (Figs. 22, 23) and became narrower cephalad as well as caudad. The lumen of the rhombencephalon, the rhombocoele, was narrowest at the level of the first branchial groove and appeared as a dorsoventrally oriented, slit-like space (Figs. 18, 19). Traced posteriorly the rhombocoele widened, and the rhombencephalon on transverse section was first V-shaped and then oval with the larger side upward and with a dorsoventrally oriented long axis.

Further posteriorly the rhombencephalon was closed dorsally with a thin pseudostratified columnar epithelium (Fig. 29), and this condition continued until the level of the fourth somite, where the rhombencephalon merged into the spinal cord without any perceivable boundary (Fig. 32).

Caudal to this level the neural tube took a round form with a dorsoventrally oriented, spindle-shaped lumen until the level of about 0.4 mm posterior to the 15th somite, where it opened again into the amniotic cavity as the posterior neuropore (Fig. 51). More posteriorly the neural groove became rapidly shallower, and at the posterior end of the body the neuroepithelium, mesenchyme, and entoderm intermingled to form a dense cell aggregation, the primitive streak (Figs. 51–57).

3.2.1.2. Neural Crest

The neural crest cells migrating from the dorsolateral extremity of the neural plate were clearly seen in the posterior portion of the midbrain (trigeminal region) (Fig. 13) and in the region anterior to the otic pit (facio-acoustic region) (Figs. 17–19). In the former, the neural crest cells constituted a somewhat dense cell aggregation attached to the dorsolateral aspect of the neural plate. In the latter they composed a large and conspicuous cell aggregation occupying the dorsomedial portion of the second branchial arch. Their cytoplasm was stained more deeply than that of surrounding mesenchymal cells. In the region of the third branchial arch three small cell clusters of neural crest cells were successively found on each side posterior to the otic pit (Figs. 23, 24, 26).

More caudad, very small cell groups consisting of several neural crest cells on transverse section were recognized at the dorsolateral corner of the neural tube at the level of each somite from the first to the seventh (Fig. 35).

3.2.1.3 Surface Ectoderm

The surface ectoderm was in general thin and consisted of a simple cuboidal or simple squamous epithelium. In the head region, especially on the surface of the first, second, and third branchial arches (Figs. 14–24), it was thick and showed a simple columnar epithelial cell arrangement, corresponding to the ectodermal ring described by O'Rahilly and Müller (1985). The ectodermal cells constituting the first and second branchial membranes were tall and columnar in shape, corresponding to the tall, columnar entodermal cells (Figs. 15–18, 22).

3.2.2. Entoderm

Of the three segments within the entodermal tract the midgut was the largest, constituting about half of its entire length. The foregut and hindgut each comprised about half of the remaining half.

3.2.2.1 Foregut

The foregut extended from the level of the posterior portion of the midbrain to that of the posterior end of the first somite, where it opened into the yolk sac at the anterior intestinal portal and passed on to the midgut (Figs. 12–28).

The most cephalic portion of the foregut was separated from the stomatodeum by a thin two-cell-layer membrane, the buccopharyngeal membrane, having three perforations (Figs. 4,12,13). Traced posteriorly the foregut widened transversely and formed the pharyngeal cavity, being compressed dorsoventrally and tapering posteriorly (Figs. 14–24). Corresponding to the first and second branchial grooves the lateral widening of the cavity attained the maximum to form the first and second branchial pouches at the dorsolateral extremity. The first pouch on each side showed a ventral invagination at the lateral edge. At each pouch the entoderm of its lateral wall was in direct contact with the surface ectoderm, establishing a two-cell-layer membrane, the branchial membrane (Figs. 15–18, 22). The third pouch was not yet recognized.

At the level of the anterior extremity of the otic placode, i.e., at about the middle portion between the first and second branchial pouches, the floor of the pharynx made a small outpocketing on the median line. This represents the thyroid diverticulum (Fig. 19).

The entodermal cells constituting the dorsal wall of the pharynx were in general flattened simple columnar, whereas those of the lateral and ventral walls were pseudostratified columnar, having two to four rows of nuclei.

From the anterior end of the foregut to the level of the second branchial pouch the notochord was embedded in the mid-dorsal line entoderm (Figs. 12–22).

At the level of the anterior intestinal portal the entoderm of the most dorsal portion of the anterior wall of the yolk sac showed a conspicuous thickening as the liver primordium, but the outgrowth of hepatic cell cords into the substance of the septum transversum was not yet noted (Figs. 26–28). The tracheo-pulmonary primordium was not identified in this specimen.

3.2.2.2 Midgut and Hindgut

The midgut extended between the levels of the 2nd–12th somites and constituted the mid-dorsal portion of the yolk sac (Figs. 28–42). At the level of the second and third somites it appeared as a somewhat deep dorsal depression in the mid-ventral portion of the embryonic body (Figs. 28–31). The entodermal cells of this portion were simple flattened columnar on the dorsal wall and pseudostratified columnar on the lateral wall, medial to the coelomic duct. Posterior to this level the midgut became shallower, and the entodermal cells became also flattened simple columnar (Figs. 32–39).

The hindgut began at the level of the 13th somite and continued until the level of about 0.4 mm posterior to the posterior neuropore (Fig. 43). The hindgut was generally oval with a dorsoventrally oriented long axis. At the level of about 0.1 mm posterior to the 15th somite the allantoic duct originated as a thin tubular structure from the ventral extremity of the hindgut, ran first ventrally and then posteriorly surrounded by the body stalk mesenchyme (Figs. 48–56).

Posterior to the origin of the allantoic duct the entoderm of the mid-ventral wall of the hindgut (cloaca) came in contact with the surface ectoderm, forming a two-cell-layer membrane, the cloacal membrane, which closed ventrally the relatively large lumen of the cloaca (Figs. 50–54).

Between the levels of the beginning of the hindgut and that of the allantois, numerous large round lucent cells with a large spherical nucleus were found among the entodermal cells lining the ventral half of the hindgut. They represented the primordial germ cells. Some of them were also found among the mesenchymal cells beneath the hindgut entoderm (Figs. 46, 47).

3.2.3 Mesoderm

3.2.3.1 Somites

As mentioned above, 15 pairs of somites were recognized in this embryo (Fig. 7). The first somites were quite small on transverse sections, and only their dorsolateral portions showed a simple columnar epithelial cell arrangement, representing the dermomyotome (Figs. 26, 27). Cells of medial and ventral portions had come loose, showing the state of the mesenchyme, i.e., sclerotome. The second somites were somewhat larger than the first, but the pattern of cell arrangement was roughly the same as the first (Figs. 28, 29).

The third, fourth, and fifth somites were large and appeared triangular in shape, consisting of dorsolateral dermomyotome and ventromedial sclerotome. In the former, cells lying beneath the surface ectoderm were disposed as a simple columnar epithelium parallel to the surface ectoderm indicating the dorsal boundary of the myocoele, the lumen of which, however, was not recognizable. In the latter, cells made a densely packed mesenchymal cell aggregation, from the ventrolateral aspect of which cells dispersed ventrolaterally (Figs. 31, 32, 34).

From the sixth to the tenth somite, somites showed a relatively well defined triangular contour, and each contained a lumen (Figs. 35–37, 39, 40), the myocoele, which was narrow in the sixth and seventh somites and quite conspicuous in the eighth, ninth, and tenth somites. In the latter three somites cells were disposed radially around the myocoele, in which a variable number of stellate cells were seen. The boundary of the myocoele was well defined only at the dorsolateral portion including the dorsomedial corner.

The last five somites gradually diminished in size and were quadrangular in shape. The cells were arranged as a pseudostratified columnar epithelium around the narrow but distinct myocoele (Figs. 41–45).

Caudal to the 15th somite, the medial mesoderm was no longer segmented and continued until the caudal end of the body, where it joined in the formation of the primitive streak (Figs. 46–57).

3.2.3.2 Intermediate Mesoderm

The intermediate mesoderm first appeared on the left side at the level of the seventh somite as a small cell group consisting of small polyhedric cells with a round lucent nucleus and a scant cytoplasm between the ventrolateral aspect of the somite and the coelomic epithelium lining the dorsomedial portion of the coelomic cavity (Fig. 36). On the right side a small cell group of this kind was seen embedded in the coelomic epithelium at the dorsomedial corner of the coelomic cavity. The latter condition continued on each side until the level of the ninth somite (Figs. 36–39).

In the region of the 10th–15th somites cells constituting this cell group increased in number, formed a conspicuous and more or less independent cell aggregation, and shifted dorsolaterally, causing a faint bulge in the surface ectoderm above it (Figs. 40–45). Posterior to the 15th somite the cell aggregation became loose and interposed again between the medial mesoderm and coelomic epithelium (Figs. 46, 47).

Indications of pronephric differentiation were noted nowhere along the whole length.

3.2.3.3. Coelomic Cavity

The coelomic cavity of the embryo was divided into four portions: the pericardiac cavity, the coelomic duct, the coelomic cavity opening laterally into the extra-embryonic coelom, and the caudal coelomic cavity lying lateral to the cloaca.

The pericardiac cavity appeared as a very large cavity extending from the level just behind the stomatodeum to that of the posterior portion of the first somite and bulging ventrally (Figs. 14–26). Due to the fact that the surface ectoderm lining the anterior and lateral aspects of the pericardiac swelling turned dorsally to become the amnion at about the middle of the aspects, the ventral aspect of the pericardiac cavity opened into the extra-embryonic coelom (Figs. 14–21). The ventral aspect of the pericardiac cavity was first closed by the septum transversum at the level of the posterior end of the otic placode (Fig. 22). In most parts the pericardiac cavity was nearly filled by the heart tube, and the cavity itself was reduced to a slit-like space. The mesodermal cells lining the pericardiac cavity were tall, spindle-shaped and appeared as a simple columnar epithelium on its dorsal wall, but they became lower traced ventrally on the lateral wall and attained the condition of simple squamous epithelium on the ventral wall.

The coelomic duct began on each side at the level of the posterior portion of the first somite where it passed on to the posterior end of the pericardiac cavity (Fig. 27), ran posteriorly as a straight duct dorsal then dorsomedial to the sinus venosus and lateral to the dorsal portion of the midgut, and joined the third portion at the level of the fifth somite (Figs. 33, 34). The lining cells were spindle-shaped and surrounded the lumen as a simple columnar epithelium. The spindle-shaped cells were dispersing radially at many places.

The third portion, the coelomic cavity opening laterally into the extra-embryonic coelom, extended between the level of the fifth somite and that of about 0.1 mm posterior to the 15th somite where the allantois orig-inated, and was closed ventrally by the loose mesenchymal tissue of the body stalk (Figs. 34–49). The lining cells of this portion showed in general the same characteristics as those of the coelomic duct. They underwent a sudden division into the loose mesenchymal tissue surrounding the yolk sac ventrolaterally and that underlining the amnion dorsolaterally (Figs. 34–38).

The caudal coelomic cavity was located on each side lateral to the cloaca (Figs. 49–53). Morphologically, the lining cells were similar to those of the coelomic duct. Cells of this portion were dispersing radially every-where, underlying the surface ectoderm and the entoderm of the cloaca.

3.2.3.4 Notochord

The notochord formed a continuous cord-like structure located between the central nervous system and the alimentary tract and extending along the median line from the anterior end of the foregut to the caudal end of the hindgut (cloaca) where this joined in forming the primitive streak (Figs. 11–55). In the region of the anterior two-thirds of the foregut its ventral portion was embedded in the entoderm on the mid-dorsal line so that its ventral surface faced the pharyngeal cavity. In most parts it showed a well-defined round contour, consisted of 5–15 small polyhedric cells on the transverse section, and stayed in contact with the mid-ventral aspect of the central nervous system.

3.2.4 Cardio-vascular System

3.2.4.1 Heart

The heart tube was relatively large, took an S-shaped tortuous course and occupied almost entirely the pericardiac cavity. It comprised bulbus cordis, ventriculus, and atrium, each of which consisted of a thick myocardial tube and a thin endocardial tube located along the axial portion of the former. The two tubes were separated from each other by a wide transparent space containing very few cellular elements and no visible coagulum (Figs. 15–26).

The most cephalic portion of the bulbus cordis was connected with the mid-dorsal aspect of the pericardiac wall at the level of the first branchial pouch, just anterior to the thyroid diverticulum, where the endocardial tube passed on to the aortic sac (Figs. 17, 18). The bulbus cordis ran first right-ventrally then ventroposteriorly to its junction with the ventriculus, which in turn went transversely to the left then posterodorsally and joined the atrium at about the level of the second branchial pouch (Figs. 15–22).

The atrium extended transversely to the right in the posterodorsal portion of the pericardiac cavity and made a posterior extension on each side (Figs. 22–26), which at the level of the first somite passed on to the sinus venosus embedded in the mesenchymal tissue of the septum transversum on each side of the anterior intestinal portal and ventral to the coelomic duct (Figs. 27–31).

In the course of the endocardial tube there was an evident constriction at the transition from the atrium to the ventriculus and a slight one between the ventriculus and the bulbus.

Abundant nucleated blood cells were seen in the endocardial lumen of the ventriculus, atrium, and sinus venosus but very few in the bulbus and aortae.

The dorsal mesocardium was clearly seen in the most cephalic region of the bulbus cordis (Figs. 20–22), but more posteriorly it underwent degeneration and only vestigeal structures were seen in the region of the atrium (Figs. 24, 25).

Except for the most cephalic portion of the bulbus cordis, where the surface of the myocardial tube was covered by the epicardial epithelial cells continuous with the pericardial epithelium (Figs. 18, 19), the myocardial tube had mostly no epicardial covering and directly faced the pericardiac cavity.

3.2.4.2 Aortae

From the aortic sac, paired ventral aortae ran cephalad on each side of the median line, ventral to the pharynx (Figs. 15–17). At the level just anterior to the first branchial pouch they turned dorsalwards, passed along the lateral wall of the pharynx forming the first branchial arch arteries, and joined the dorsal aortae dorsal to the pharynx (Figs. 12–14).

The dorsal aortae ran straight caudalwards and were throughout their course situated on each side of the notochord and closely applied to the dorsal wall of the alimentary tract (Figs. 14–52). They extended caudally to the level of the posterior one-third of the cloaca where they became very thin and were no longer traced as continuous vessels (Figs. 53, 54). No transverse anastomosis was noted between the right and left aortae throughout their course. There was no sign of the formation of the second branchial arch artery.

From the dorsal aortae sproutings of the dorsal intersegmental arteries were identified at the levels of the eighth, ninth, and tenth intersegmental spaces (Fig. 38), and that of the lateral segmental arteries were suggested with a mesenchymal cell cord at the levels of the eighth and ninth somites. The ventral segmental arteries arose at the levels of each somite from the 7th to the 15th somite and ran ventrally along the lateral aspect of the yolk sac or hindgut (Figs. 41, 44, 49).

More posteriorly, in the region of the body stalk, three pairs of ventral segmental arteries were noted. Their endothelial lining was, however, not clear at their ventral end, where the endothelial cells intermingled with mesenchymal cells of the body stalk (Fig. 49). For this reason, the direct continuation between these ventral segmental arteries, and a single large blood vessel, the umbilical artery, in the body stalk was nowhere recognized.

In the head region, numerous blood vessels of relatively wide caliber, presumably the primordium of the cephalic veins, were seen directly beneath the neuroepithelium of the brain, but their relationship with the dorsal aortae could not be identified (Figs. 13, 14, 16, 17).

3.2.4.3 Umbilical Veins

The umbilical veins began at the dorsolateral extremity of the sinus venosus at the level of the third somite (Fig. 31) and ran straight caudalwards in a fold of the somatopleura which protruded into the exocoelomic cavity along the line of the amniotic attachment (Figs. 33–46). Posterior to the level of the 14th somite they turned ventrally with the somatopleura, and at the level of the beginning of the allantois the mesenchymal cells surrounding the umbilical vein joined with that of the body stalk, where the endothelial lining of the umbilical vein became loose and the endothelial cells intermingled with surrounding mesenchymal cells (Fig. 49). Thus, the continuation between the umbilical veins and vessels in the body stalk was not recognized (Figs. 50–57).

4 Discussion

The embryo described above had 15 pairs of somites and, therefore, belongs to Streeter's developmental horizon XI (Carnegie stage 11), comprising the embryos with 13–20 pairs of somites. Streeter (1942) first made a review of horizon XI based on 16 embryos. Recently Müller and O'Rahilly (1986) surveyed 20 embryos belonging to stage 11 with special reference to the development of the central nervous system, 15 of which were once dealt with in Streeter's review. O'Rahilly and Müller (1987) further published a comprehensive study on human embryos from stage 2 to stage 23, in which 23 embryos belonging to stage 11 were investigated.

Among the embryos reported in the literature, the following were in a stage of development comparable to that of our embryo: embryo C. C. 6344 with 13 somites (Streeter 1942), embryo Pfannenstiel III with 13 or 14 somites (Low 1908), embryo C. C. 4529 with 14 somites (Heuser 1930), embryo H. 810 or Dorland's embryo with 15 somites (Dorland and Bartelmez 1922), embryo C. C. 470 with 16 somites (Bartelmez and Evans 1926), embryo C. C. 7611 with 16 somites (Streeter 1942), and embryo H. 52 or C. C. 5072 with 17 somites (Atwell 1930).

Our embryo shows in general more similarity to the younger embryos (13 or 14 somites) than to the older group (16 or 17 somites), and lies nearest the 14-somite embryo described by Heuser. Noteworthy differences between our embryo and those mentioned above are as follows:

1. Closure of the neural tube. One of the most outstanding morphological features of our embryo was the retarded closure of the neural tube. Streeter (1942) stated that "the closure of the neural tube in the younger specimens of the present age group is completed forward to the colliculi." In fact it is completed to the caudal limit of the midbrain in Heuser's 14-somite embryo. In our embryo, however, it was just completed cephalically to the level of the anterior one-third of the rhombencephalon, where the facio-acoustic neural crest cell aggregation occurred.

2. Tegmen rhombencephali. In our embryo the rhombencephalon was closed dorsally with a thin two-cell-layer membrane, the tegmen rhomben-cephali, from the level of the facio-acoustic neural crest cell aggregation to that of the first somite. This conspicuous condition has been described only in the report of a 20-somite embryo by Davis (1923); it is not mentioned in the reports cited above. Even in the detailed survey by Müller and O'Rahilly (1986) this topic was not discussed.

3. Occlusion of the neural tube lumen. Müller and O'Rahilly (1986) described the occlusion of the neural tube lumen in four of 20 embryos of this stage. In our embryo such an occlusion was found nowhere throughout the whole length of the neural tube.

4. Asymmetry of the embryonic body. Bartelmez and Evans (1926) emphasized that "so long as the nervous system dominates the external view, the human embryo appears asymmetric, sometimes strikingly so. This is true of all embryos we have studied up to the 14-somite stage, and it is due largely to differences in the shape of the two neural folds." This state, however, was not recognized in our case. As is seen in Figs. 7–32, the embryo showed practically complete symmetry.

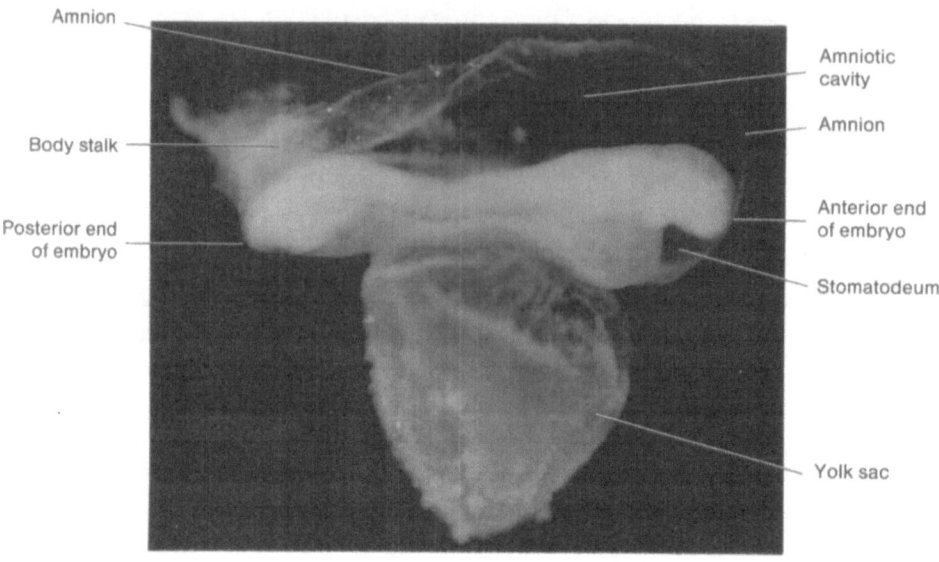

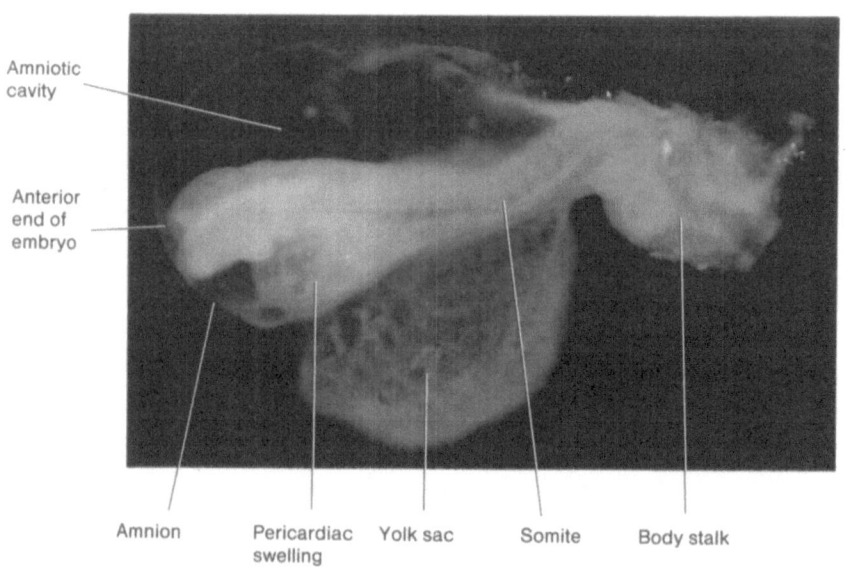

Fig. 1. (*above*) Lateral view of right side of embryo through amnion, × 16

Fig. 2. (*below*) Dorsolateral view of left side of embryo through amnion. General features of the embryo are shown. The embryonic body lies on a large yolk sac which is covered by a meshwork of highly developed blood vessels. The amniotic sac is relatively small, consisting of a thin transparent amnion through which the embryonic body is clearly seen. Between the most cephalic portion of the body and the prominent pericardiac swelling there is a distinct hollow stomatodeum. The body stalk arises from the ventral aspect of the most caudal portion of the body, runs first ventralwards and turns then to the left, × 18

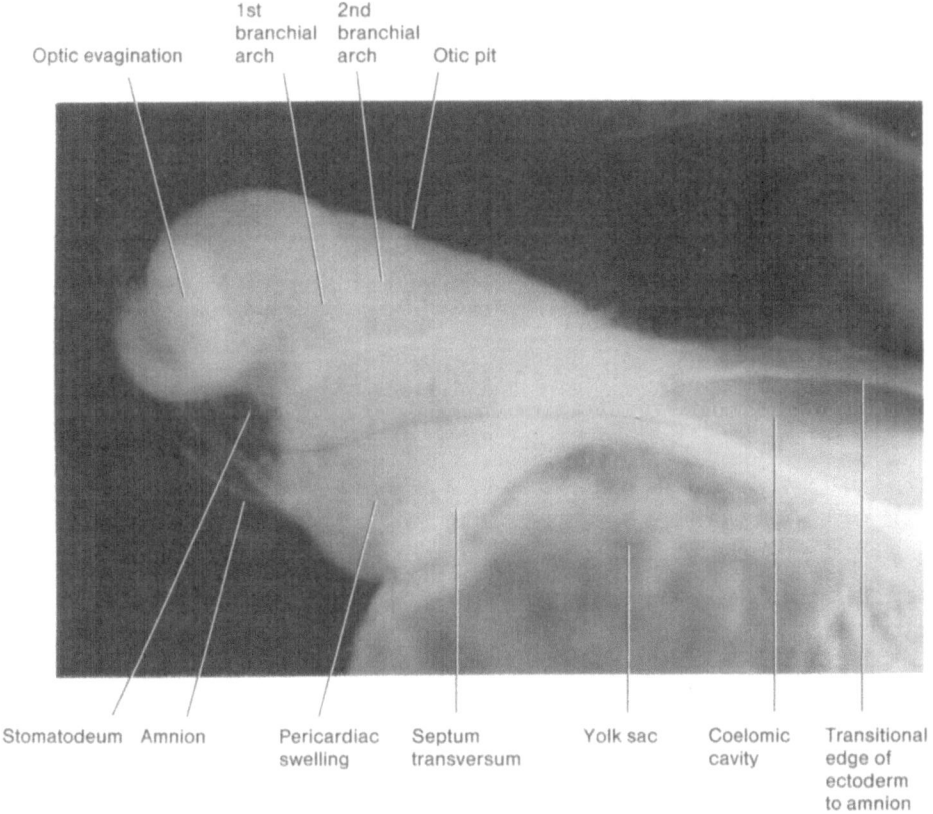

Optic evagination · 1st branchial arch · 2nd branchial arch · Otic pit

Stomatodeum · Amnion · Pericardiac swelling · Septum transversum · Yolk sac · Coelomic cavity · Transitional edge of ectoderm to amnion

Fig. 3. Lateral view of head region of left side. Posterior to the conspicuous optic evagination the first and second branchial arches follow successively. The ventral half of the head region consists of a prominent pericardiac swelling which is connected with the yolk sac by the marked septum transversum. The coelomic cavity widely opens laterally into the extra-embryonic coelom, × 36

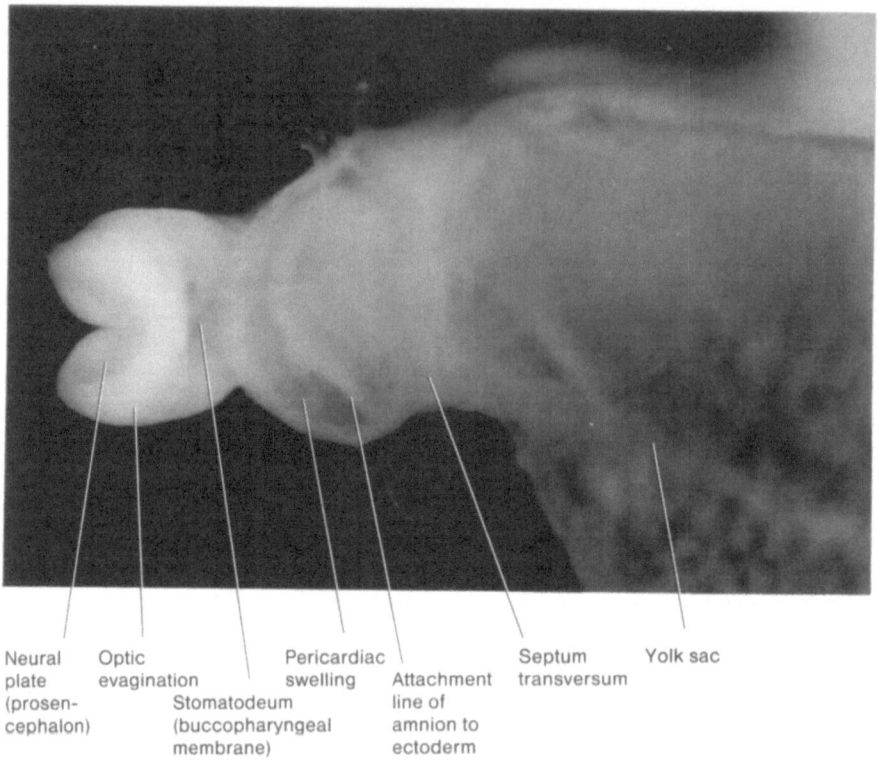

Neural | Optic | | Pericardiac | | Septum | Yolk sac
plate | evagination | | swelling | Attachment | transversum
(prosen- | | Stomatodeum | | line of
cephalon) | | (buccopharyngeal | | amnion to
 | | membrane) | | ectoderm

Fig. 4. Ventral view of head region. Posterior to the prosencephalon consisting of the optic evagination on each side the stomatodeum is evident. It is floored by the buccopharyngeal membrane, on which three perforations are perceivable. At the middle portion of the ventral and lateral aspects of the pericardiac swelling an attachment line of the amnion to the surface ectoderm is clearly seen, × 36

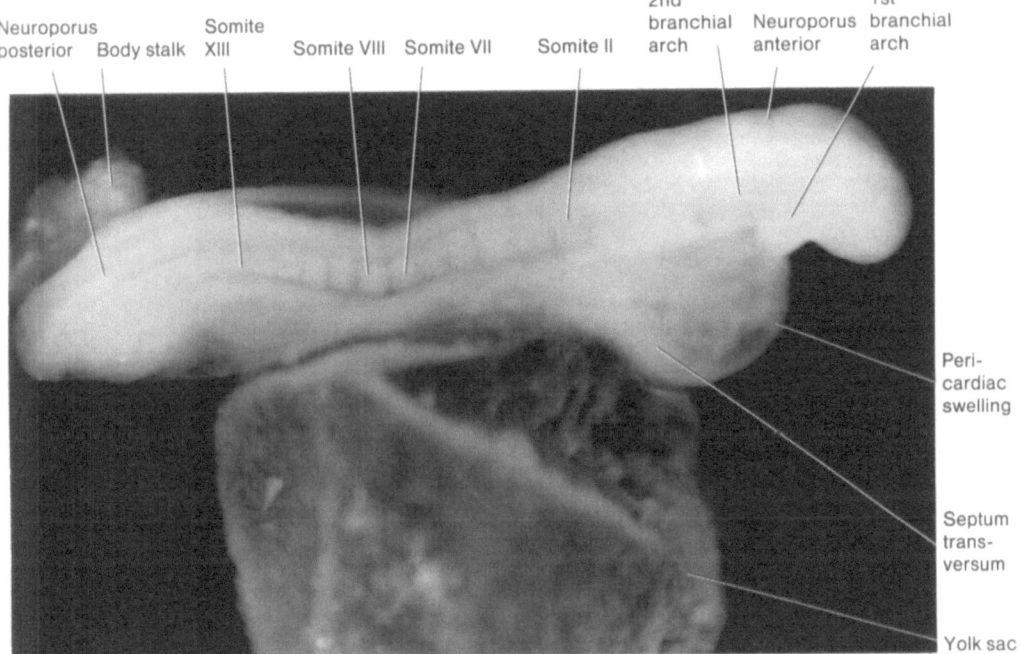

Neuroporus posterior · Body stalk · Somite XIII · Somite VIII · Somite VII · Somite II · 2nd branchial arch · Neuroporus anterior · 1st branchial arch · Pericardiac swelling · Septum transversum · Yolk sac

Fig. 5. Dorsolateral view of right side. General features of the embryo are shown in detail. The dorsal flexure centered over the seventh and eighth somites is evident. The posterior neuropore is also noted, × 29

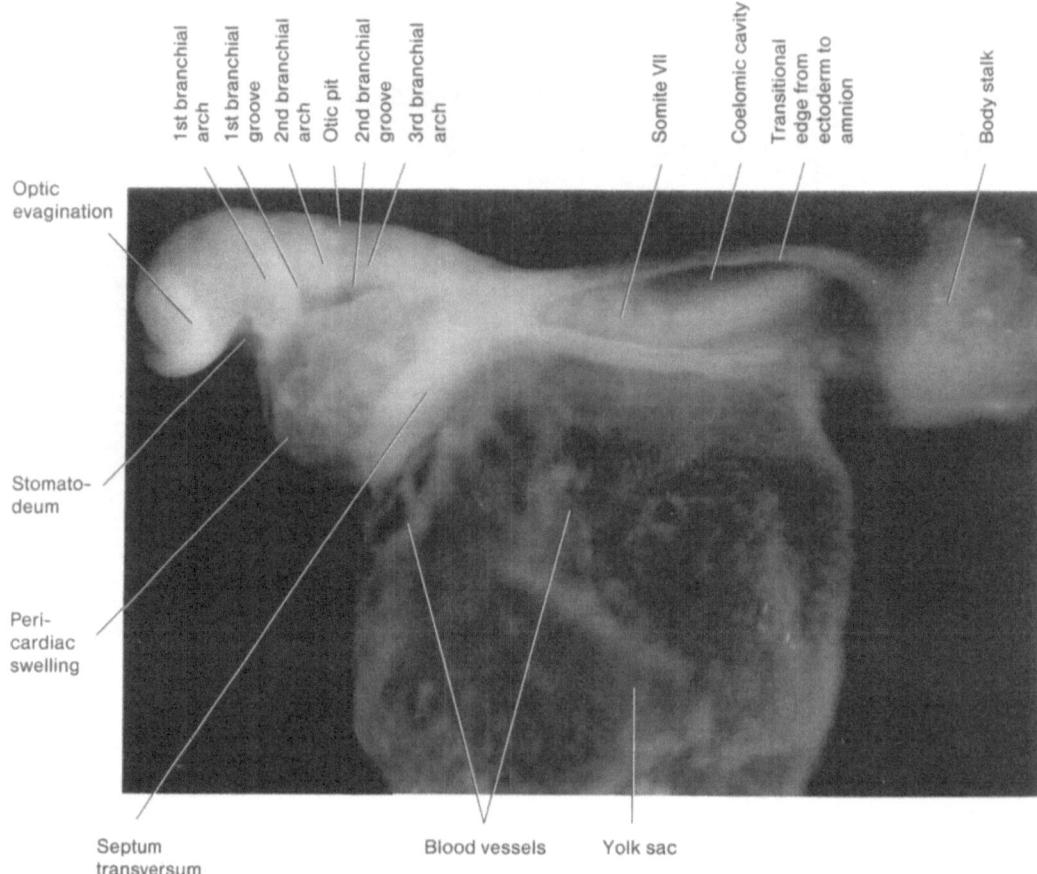

Optic evagination

1st branchial arch
1st branchial groove
2nd branchial arch
Otic pit
2nd branchial groove
3rd branchial arch

Somite VII

Coelomic cavity

Transitional edge from ectoderm to amnion

Body stalk

Stomato-deum

Peri-cardiac swelling

Septum transversum

Blood vessels

Yolk sac

Fig. 6. Ventrolateral view of left side. Characteristic features of the cephalic portion are clearly seen: optic evagination, the first and second branchial arches, the first and second branchial grooves, otic pit, stomatodeum, pericardiac swelling, septum transversum, and yolk sac. In the posterior half of the embryo the lateral opening of the coelomic cavity into the extra-embryonic coelom is evident, ×29

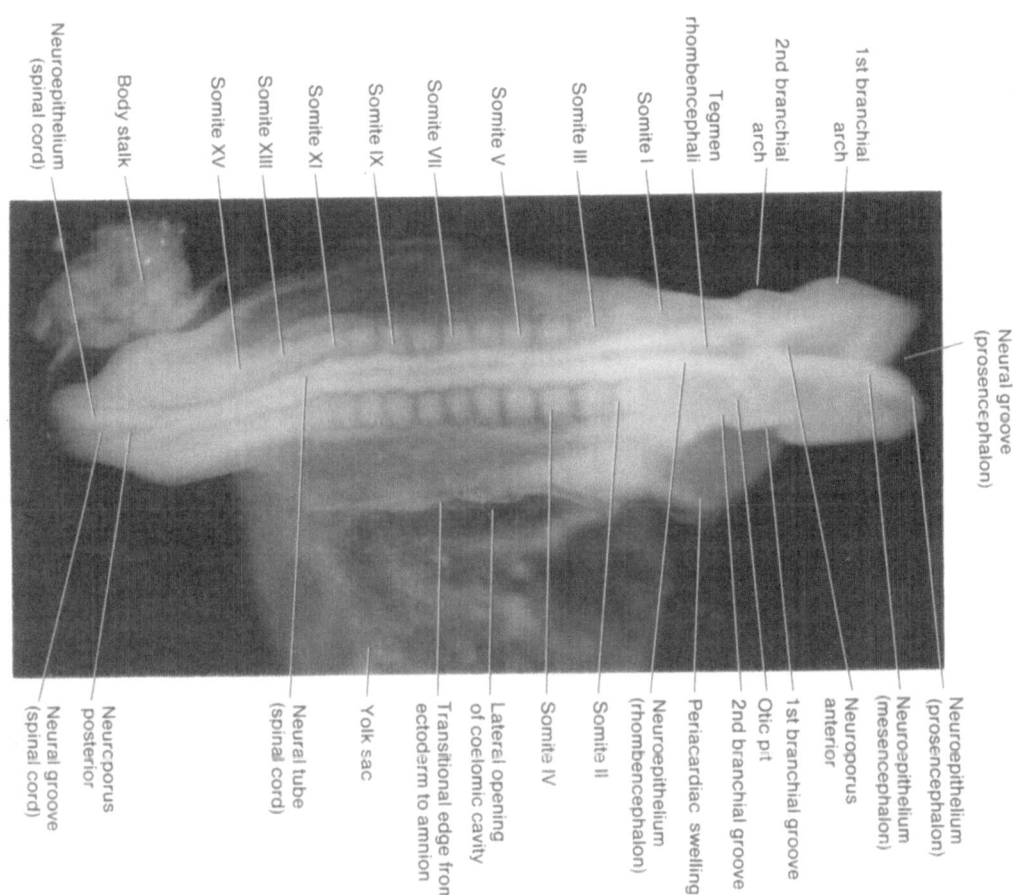

Neural groove (prosencephalon)

Neuroepithelium (spinal cord)

Body stalk

Somite XV

Somite XIII

Somite XI

Somite IX

Somite VII

Somite V

Somite III

Somite I

Tegmen rhombencephali

2nd branchial arch

1st branchial arch

1st branchial arch

2nd branchial arch

Neural groove (spinal cord)

Neuroporus posterior

Neural tube (spinal cord)

Yolk sac

Transitional edge from ectoderm to amnion

Lateral opening of coelomic cavity

Somite IV

Somite II

Neuroepithelium (rhombencephalon)

Periacardiac swelling

2nd branchial groove

Otic pit

1st branchial groove

Neuroporus anterior

Neuroepithelium (mesencephalon)

Neuroepithelium (prosencephalon)

Fig. 7. Dorsal view. The anterior and posterior neuropores, neural tube, and the symmetrical arrangement of somites are clearly visible. The first as well as the 15th somites are only faintly indicated. The tegmen rhombencephali is also faintly perceivable, × 29

19

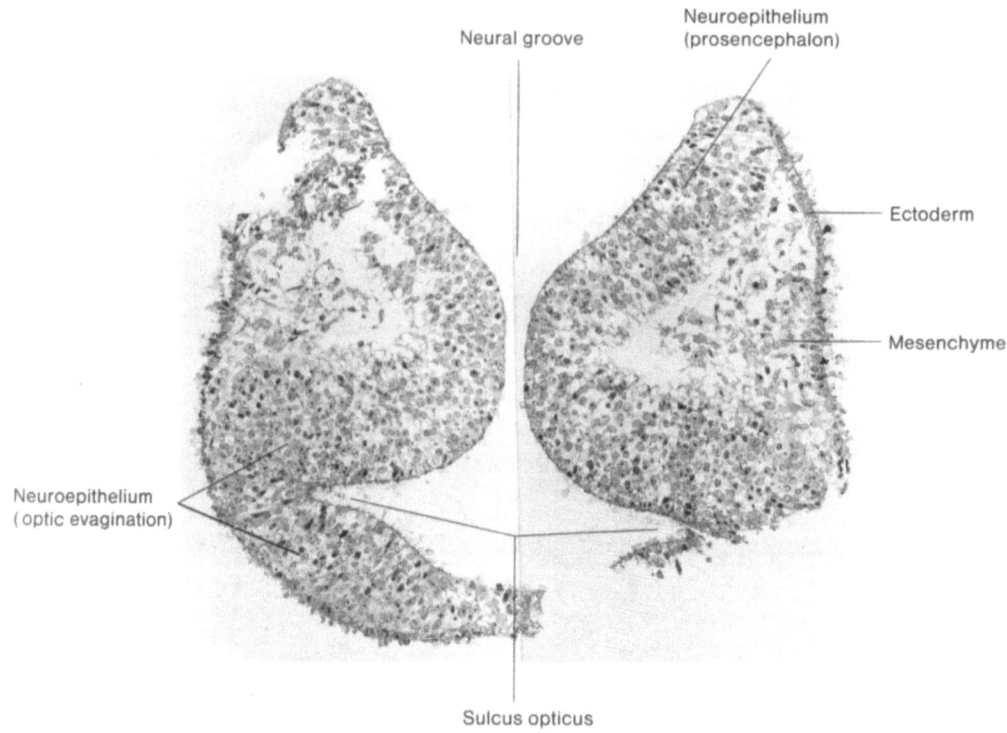

Fig. 8. Section through anterior end of embryo (3–1). The ventral half of the neural plate evaginates laterally and constitutes a conspicuous optic evagination in which optic sulcus is evident. The dorsal half of the neural plate represents the prosencephalon (diencephalon) and consists of a tall pseudostratified columnar epithelium. The surface ectoderm consists in general of a simple columnar epithelium and shows no sign of differentiation into the lens placode, although it is in direct contact with the ventral and ventrolateral aspects of the optic evagination

Figures 8–57 are photomicrographs of transverse sections of the embryo stained with toluidine blue. The numbers in parentheses are the serial numbers of the sections. Magnification: Figs. 8–19 and 44–57, × 120; Figs. 20–43, × 150

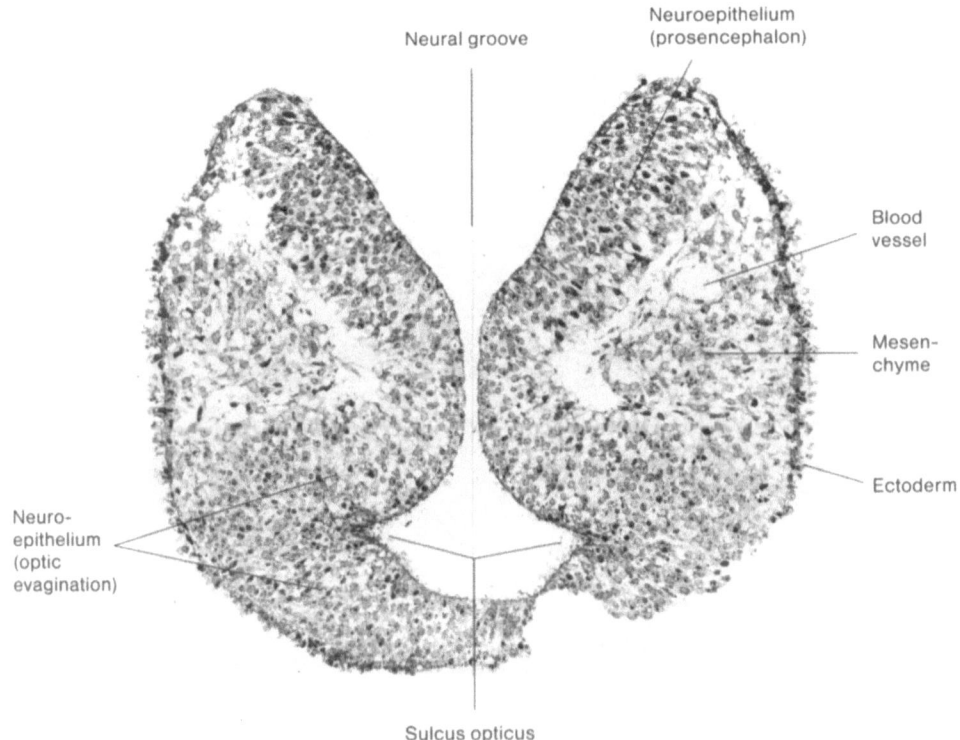

Neural groove

Neuroepithelium
(prosencephalon)

Blood
vessel

Mesen-
chyme

Ectoderm

Neuro-
epithelium
(optic
evagination)

Sulcus opticus

Fig. 9. Section through middle portion of optic evagination (3–2). The morphological features are generally similar to those seen in Fig. 8, except for increased mesenchymal cells between the neural plate and surface ectoderm. Among the mesenchymal cells several blood vessels consisting exclusively of endothelial cells are seen

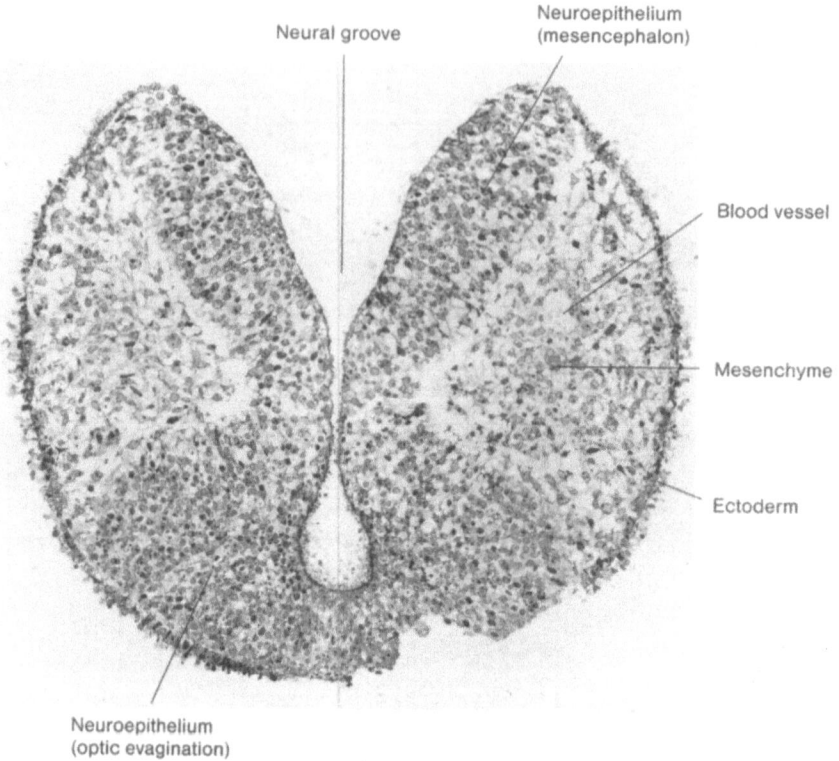

Fig. 10. Section through posterior portion of optic evagination (3–3). The neural plate dorsal to the optic evagination is now the mesencephalon. Increased mesenchymal cells cause a lateral bulge of the body surface

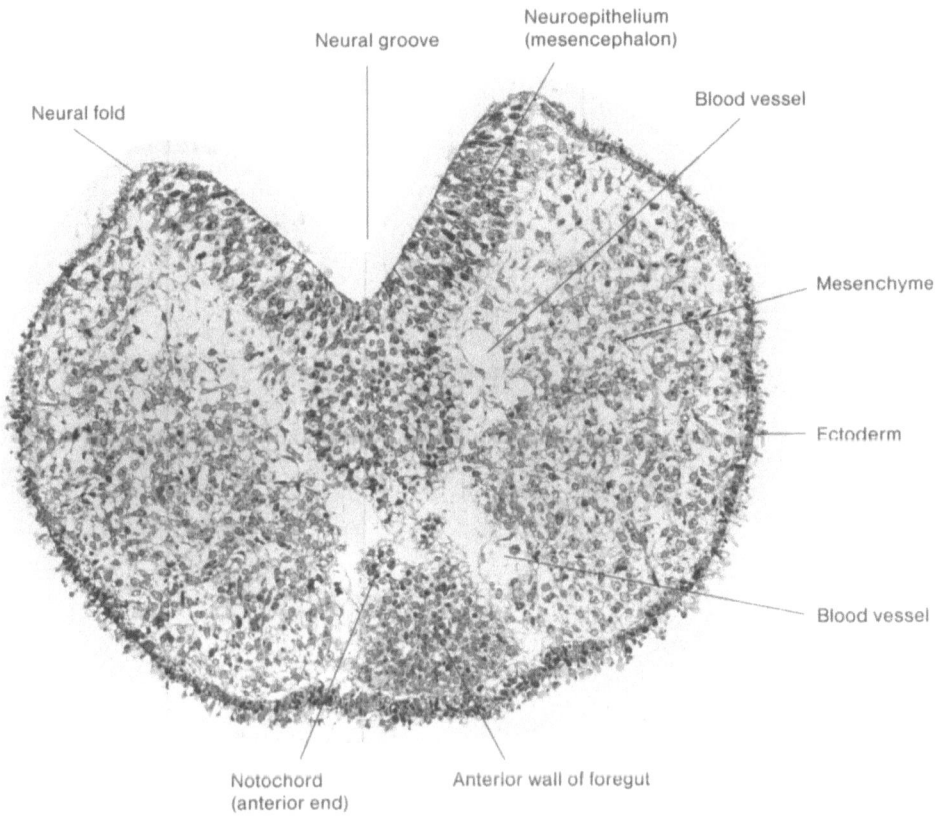

Neural groove

Neuroepithelium
(mesencephalon)

Neural fold

Blood vessel

Mesenchyme

Ectoderm

Blood vessel

Notochord
(anterior end)

Anterior wall of foregut

Fig. 11. Section through anterior end of notochord (4). In the ventrocentral part of the section dorsal to the surface ectoderm there is a cell aggregation representing the anterior wall of the foregut, and just dorsal to it an oval contour of the notochord is seen. The neural plate consisting of six to eight rows of nuclei represents the mesencephalon and shows a widely opened V-shape. Among the increased mesenchymal cells several blood vessels are recognized

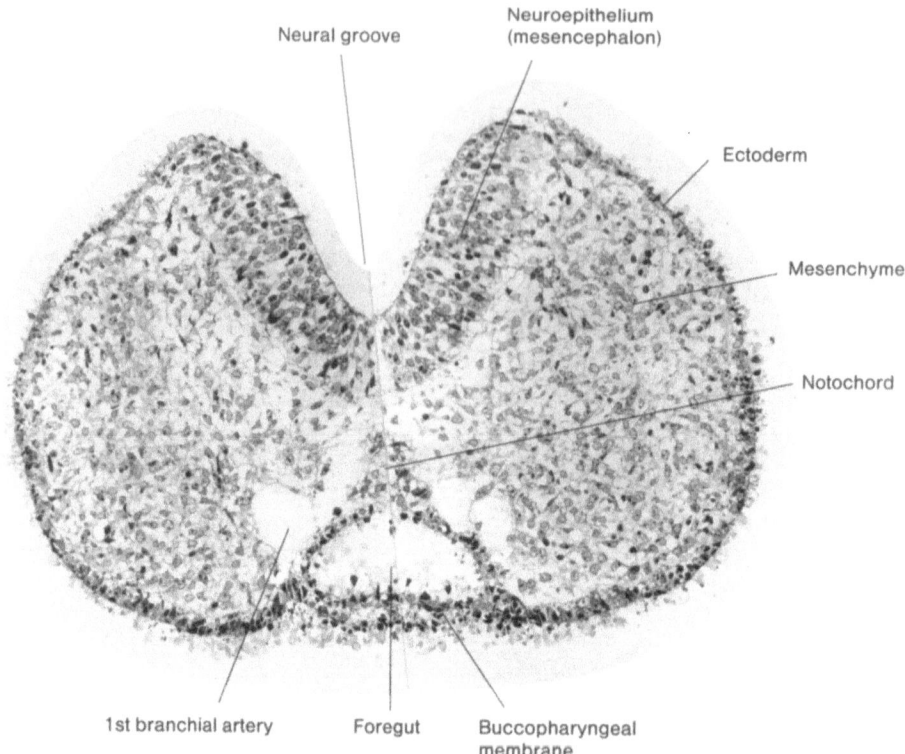

Neural groove

Neuroepithelium
(mesencephalon)

Ectoderm

Mesenchyme

Notochord

1st branchial artery   Foregut   Buccopharyngeal
membrane

Fig. 12. Section through anterior end of foregut (5). In the ventrocentral part of the section a small lumen of the foregut is seen whose ventral aspect is closed with a two-cell-layer membrane, the buccopharyngeal membrane. On the mid-dorsal wall of the foregut the notochord is embedded in the entoderm so that its ventral surface faces the foregut cavity. The mesenchymal cells filling the space between the neural plate (mesencephalon) and surface ectoderm increase greatly in number and cause a conspicuous lateral bulge of the body surface, indicating the beginning of the first branchial arch. Dorsolateral to the foregut a blood vessel of large caliber is seen on each side; this is the first branchial arch artery

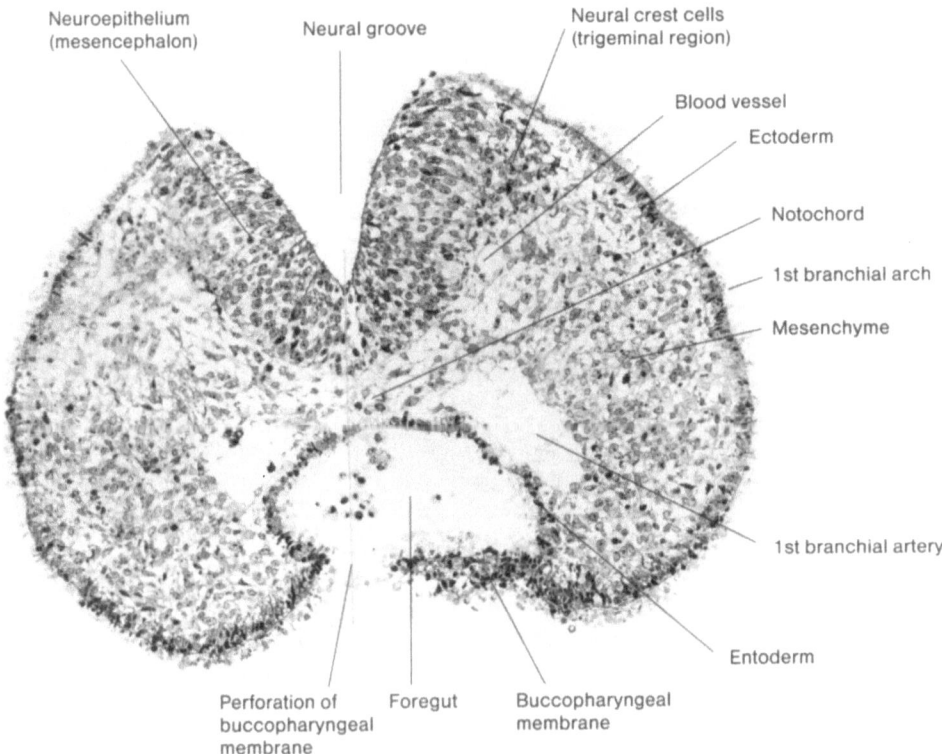

Neuroepithelium (mesencephalon)

Neural groove

Neural crest cells (trigeminal region)

Blood vessel

Ectoderm

Notochord

1st branchial arch

Mesenchyme

1st branchial artery

Entoderm

Perforation of buccopharyngeal membrane

Foregut

Buccopharyngeal membrane

Fig. 13. Section through first branchial arch artery and trigeminal neural crest region (6–3). The lumen of the foregut enlarges laterally as well as dorsally. Its ventral aspect is closed by the buccopharyngeal membrane, which has an evident perforation. Dorsolateral to the foregut the first branchial arch artery is marked on each side. Lateral to the dorsolateral extremity of the neural plate, representing now the cephalic end of the rhombencephalon, the trigeminal neural crest cell aggregation is evident

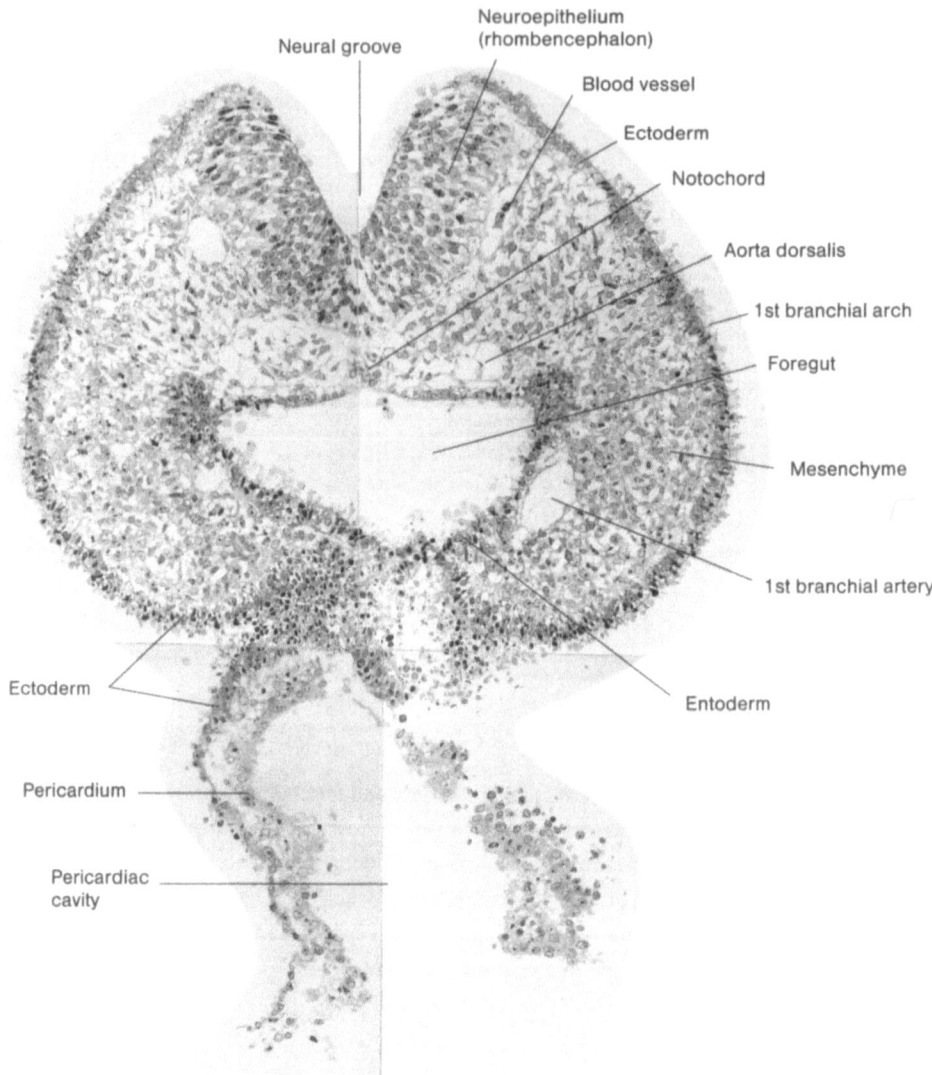

Neural groove

Neuroepithelium (rhombencephalon)

Blood vessel

Ectoderm

Notochord

Aorta dorsalis

1st branchial arch

Foregut

Mesenchyme

1st branchial artery

Ectoderm

Entoderm

Pericardium

Pericardiac cavity

Fig. 14. Section through anterior end of pericardium (7–2). The lateral bulge of the body surface, caused by the increased mesenchymal cells, is conspicuous; this indicates the mandibular arch. Symmetrical arrangement of the dorsal aortae and the first branchial arch arteries is clearly seen on the dorsal and ventrolateral aspects of the foregut respectively. The neural plate, the rhombencephalon, shows a V-shaped form and widely opens dorsally

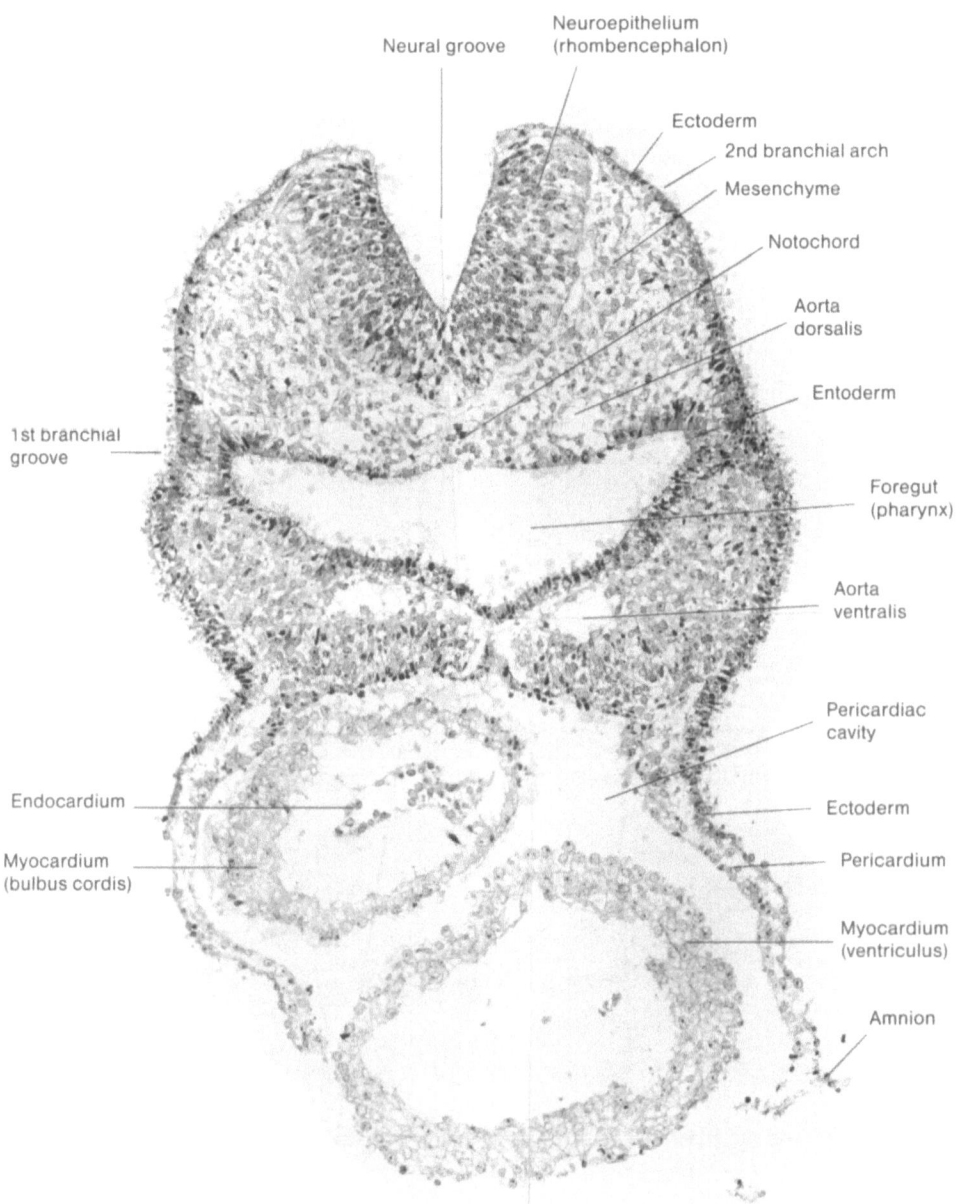

Neural groove

Neuroepithelium
(rhombencephalon)

Ectoderm

2nd branchial arch

Mesenchyme

Notochord

Aorta
dorsalis

Entoderm

1st branchial
groove

Foregut
(pharynx)

Aorta
ventralis

Pericardiac
cavity

Endocardium

Ectoderm

Myocardium
(bulbus cordis)

Pericardium

Myocardium
(ventriculus)

Amnion

Fig. 15. Section through anterior end of bulbus cordis and ventral aorta (8–1). The foregut noticeably widens transversely, indicating the beginning of the first branchial pouch, which corresponds to a depression of the body surface, the first branchial groove. The dorsolateral bulge dorsal to this groove indicates the second branchial arch. Dorsal and ventral to the foregut dorsal and ventral aortae are seen respectively. The ventral portion of the notochord is embedded in the entoderm of the mid-dorsal wall of the foregut. The bulbus cordis consists of endocardium and myocardium, whereas the ventriculus still of only myocardium. Pericardiac cavity opens ventrally into the extra-embryonic coelom

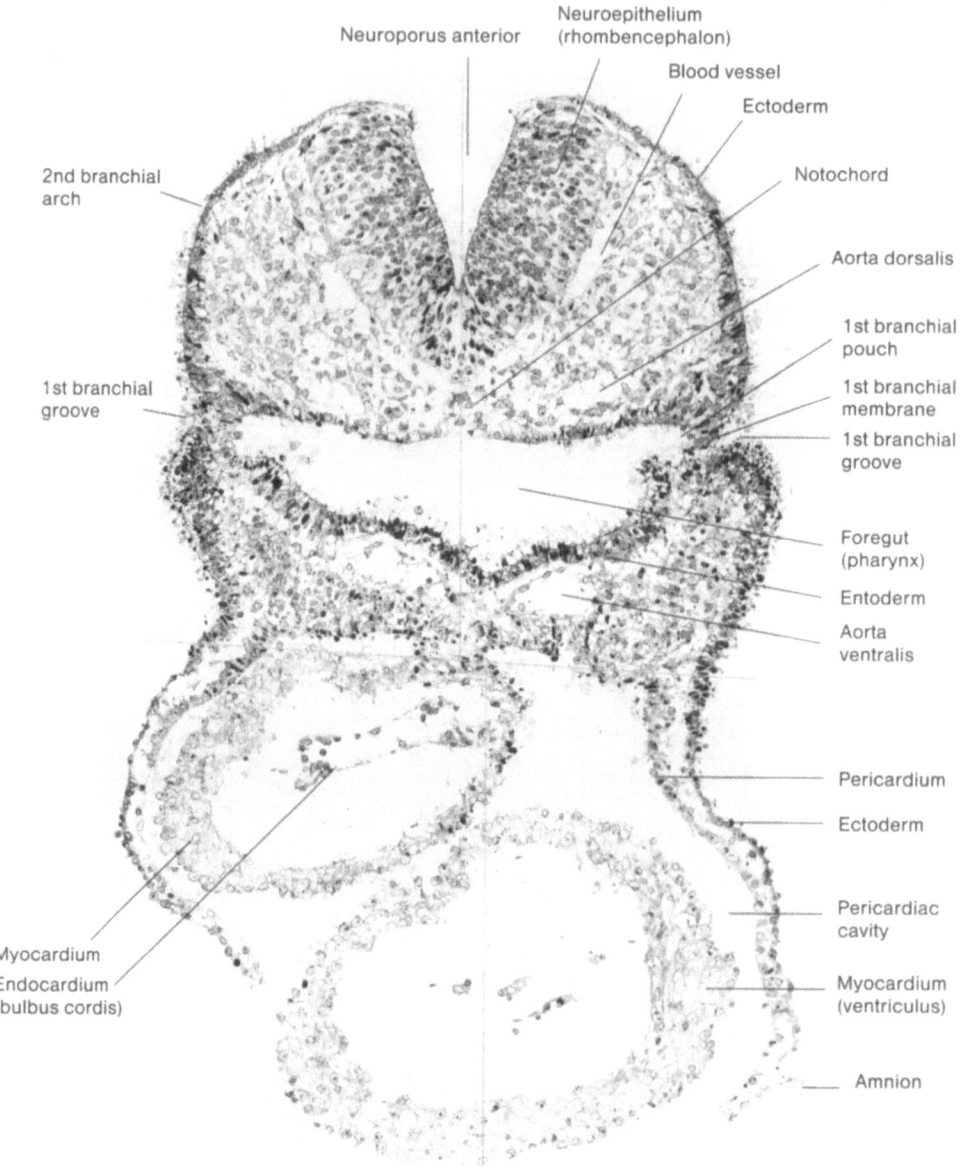

Neuroporus anterior

Neuroepithelium
(rhombencephalon)

Blood vessel

Ectoderm

2nd branchial
arch

Notochord

Aorta dorsalis

1st branchial
pouch

1st branchial
groove

1st branchial
membrane

1st branchial
groove

Foregut
(pharynx)

Entoderm

Aorta
ventralis

Pericardium

Ectoderm

Pericardiac
cavity

Myocardium
Endocardium
(bulbus cordis)

Myocardium
(ventriculus)

Amnion

Fig. 16. Section through first branchial groove and first branchial pouch (8–2). From the dorsomedial extremity of the neural plate, the rhombencephalon, sprouts a tiny medial process indicating the formation of the tegmen rhombencephali. The lumen of the foregut markedly widens transversely, forming the first branchial pouch, which is separated from the first branchial groove by a two-cell-layer membrane, the branchial membrane. Dorsal and ventral to the foregut dorsal and ventral aortae are seen respectively on each side of the median line. The appearance of the bulbus cordis, ventriculus and pericardiac cavity is quite similar to that in Fig. 15

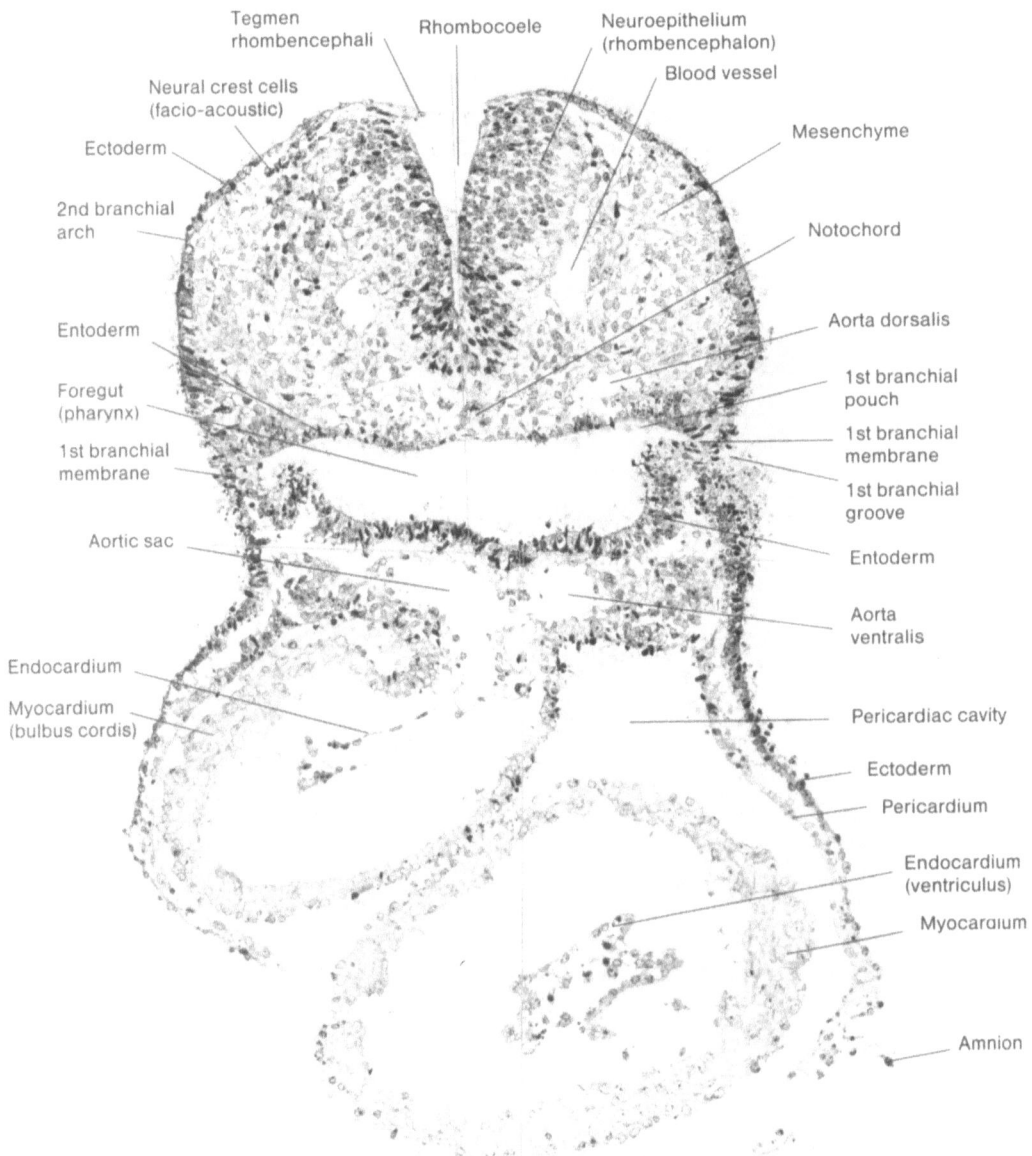

Tegmen
rhombencephali

Rhombocoele

Neuroepithelium
(rhombencephalon)

Blood vessel

Neural crest cells
(facio-acoustic)

Mesenchyme

Ectoderm

2nd branchial
arch

Notochord

Entoderm

Aorta dorsalis

1st branchial
pouch

Foregut
(pharynx)

1st branchial
membrane

1st branchial
membrane

1st branchial
groove

Aortic sac

Entoderm

Aorta
ventralis

Endocardium

Myocardium
(bulbus cordis)

Pericardiac cavity

Ectoderm

Pericardium

Endocardium
(ventriculus)

Myocaraıum

Amnion

Fig. 17. Section through anterior end of aortic sac and first branchial groove (8–3). The right and left neural plates approach each other so that the interspace displays a vertically oriented slit-like form, which opens still dorsally into the amniotic cavity. The transverse widening of the foregut is now maximal, and the first branchial pouch and corresponding first branchial groove are evident on each side. The bulbus cordis now connects with the dorsal wall of the pericardiac cavity, and the lumen of its endocardial tube opens into that of the aortic sac. The ventriculus enlarges markedly in size and now consists of myocardium and endocardium

29

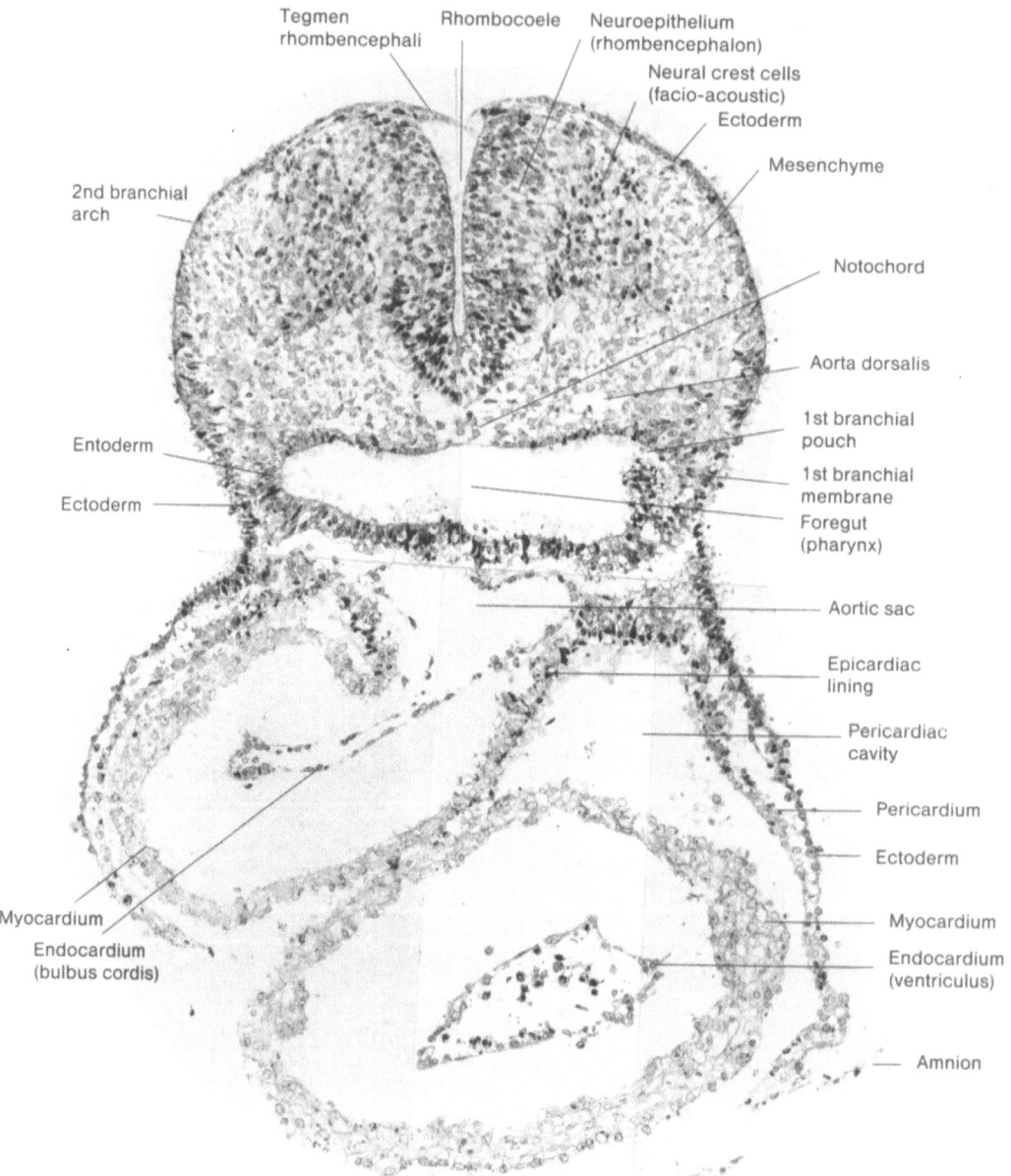

Tegmen
rhombencephali

Rhombocoele

Neuroepithelium
(rhombencephalon)

Neural crest cells
(facio-acoustic)

Ectoderm

Mesenchyme

2nd branchial
arch

Notochord

Aorta dorsalis

Entoderm

Ectoderm

1st branchial
pouch

1st branchial
membrane

Foregut
(pharynx)

Aortic sac

Epicardiac
lining

Pericardiac
cavity

Pericardium

Ectoderm

Myocardium

Endocardium
(bulbus cordis)

Myocardium

Endocardium
(ventriculus)

Amnion

Fig. 18. Section through aortic sac and facio-acoustic neural crest region (9). The right and left walls of the rhombencephalon are in close apposition and the slit-like intersapce is closed dorsally by a very thin membrane, the tegmen rhombencephali, so that the rhombo-coele is now established as a closed lumen. Lateral to the dorsal half of the rhombence-phalon the facio-acoustic neural crest cell aggregation is marked on each side. On the surface of the myocardium of the bulbus cordis and ventriculus the epicardiac lining does not exist, except that the most dorsal small area of the bulbus cordis is covered by epicardiac epithelial cells which are continuous with the pericardiac epithelium

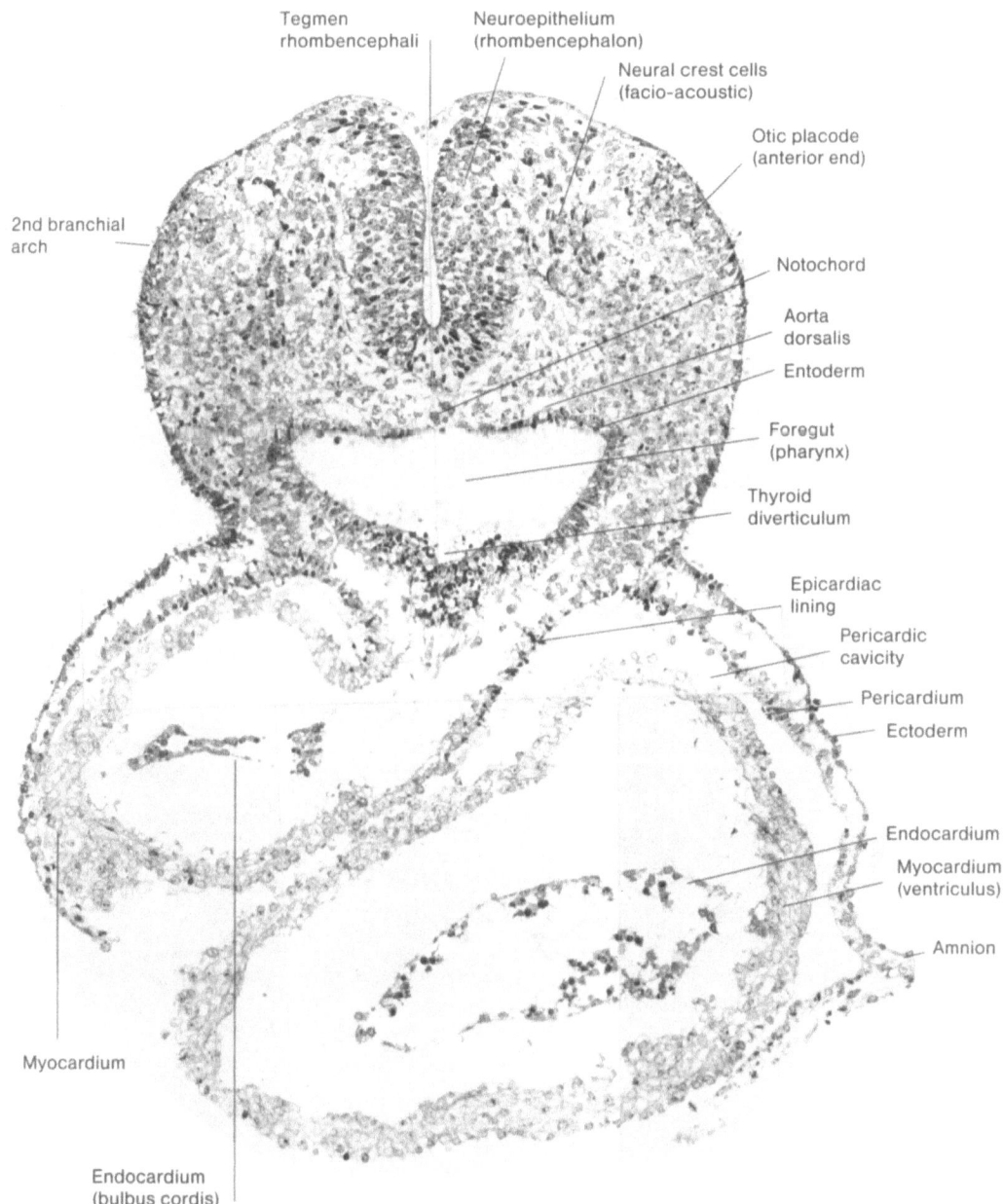

Tegmen rhombencephali

Neuroepithelium (rhombencephalon)

Neural crest cells (facio-acoustic)

Otic placode (anterior end)

2nd branchial arch

Notochord

Aorta dorsalis

Entoderm

Foregut (pharynx)

Thyroid diverticulum

Epicardiac lining

Pericardic cavicity

Pericardium

Ectoderm

Endocardium

Myocardium (ventriculus)

Amnion

Myocardium

Endocardium (bulbus cordis)

Fig. 19. Section through thyroid primordium and anterior end of otic placode (9–4). At the dorsolateral extremity of the neural plate, migrating neural crest cells are evident and an epithelial cell aggregation at the dorsolateral corner of the second branchial arch indicates the beginning of the otic placode. The notochord touches the neural plate at its mid-ventral aspect and is embedded in the entoderm of the mid-dorsal wall of the foregut. The ventral wall of the foregut, the pharynx, forms a small ventral diverticulum representing the thyroid primordium. The ventriculus is now quite large and its endocardial tube contains several nucleated blood cells. The endocardial and myocardial tubes are separated from each other by a wide transparent space containing no visible coagulum. The ventral aspect of the pericardiac cavity opens widely into the extra-embryonic coelom

Fig. 20. Section through middle portion of otic pit (10). The dorsal portion of the rhombocoele widens transversely so that the rhombencephalon itself takes a V-shaped form. Closely adjoining the dorso-lateral aspect of the rhombencephalon there is a marked otic placode consisting of a tall columnar epithelium with several rows of nuclei. In the pericardiac cavity the dorsal mesocardium is seen, connecting the mid-dorsal wall of the pericardiac cavity and the medial aspect of the bulbus cordis. The ventriculus enlarges distinctly in a left-dorsal direction. The epicardiac lining cannot be recognized either on the bulbus cordis or the ventriculus

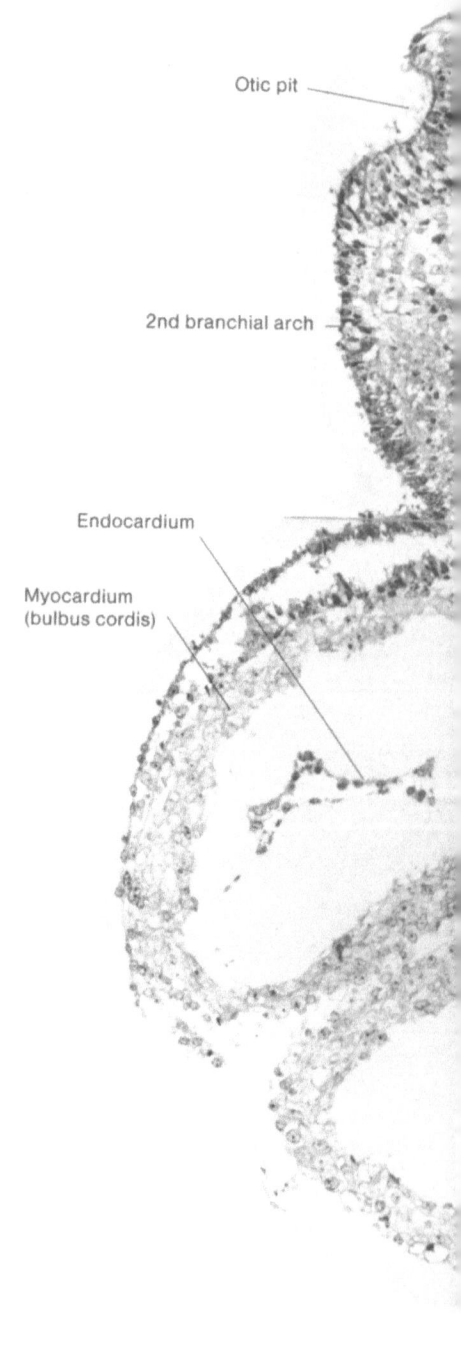

Otic pit

2nd branchial arch

Endocardium

Myocardium
(bulbus cordis)

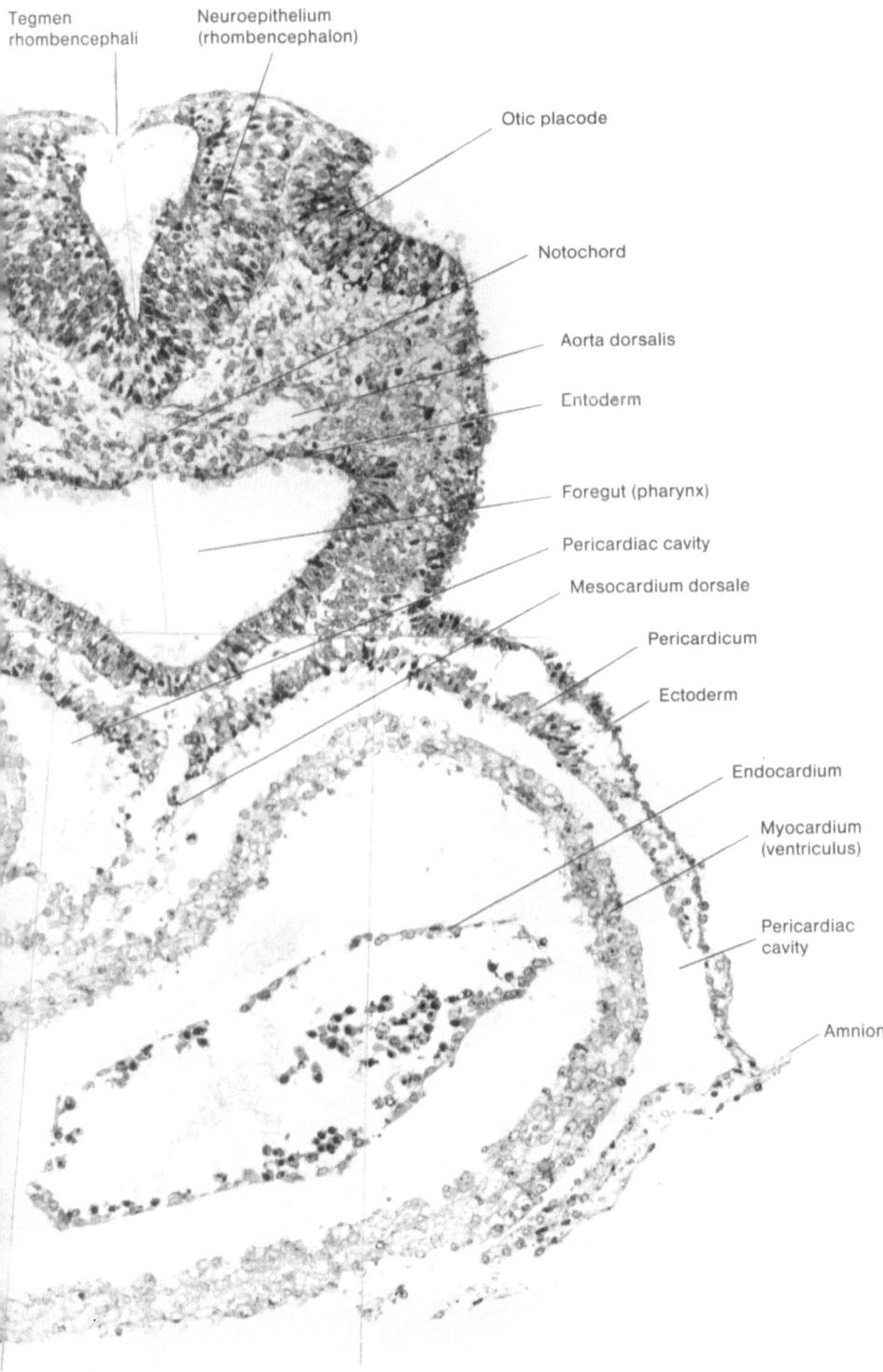

Tegmen rhombencephali

Neuroepithelium (rhombencephalon)

Otic placode

Notochord

Aorta dorsalis

Entoderm

Foregut (pharynx)

Pericardiac cavity

Mesocardium dorsale

Pericardicum

Ectoderm

Endocardium

Myocardium (ventriculus)

Pericardiac cavity

Amnion

Fig. 21. Section through posterior portion of otic pit and anterior end of atrium (11). The morphological features of the dorsal half of the body dorsal to the heart are essentially very similar to those seen in Fig. 20. The heart is now quite large and in the left-dorsal portion of the ventriculus the endocardial tube of the atrium appears. The dorsal mesocardium is degenerating

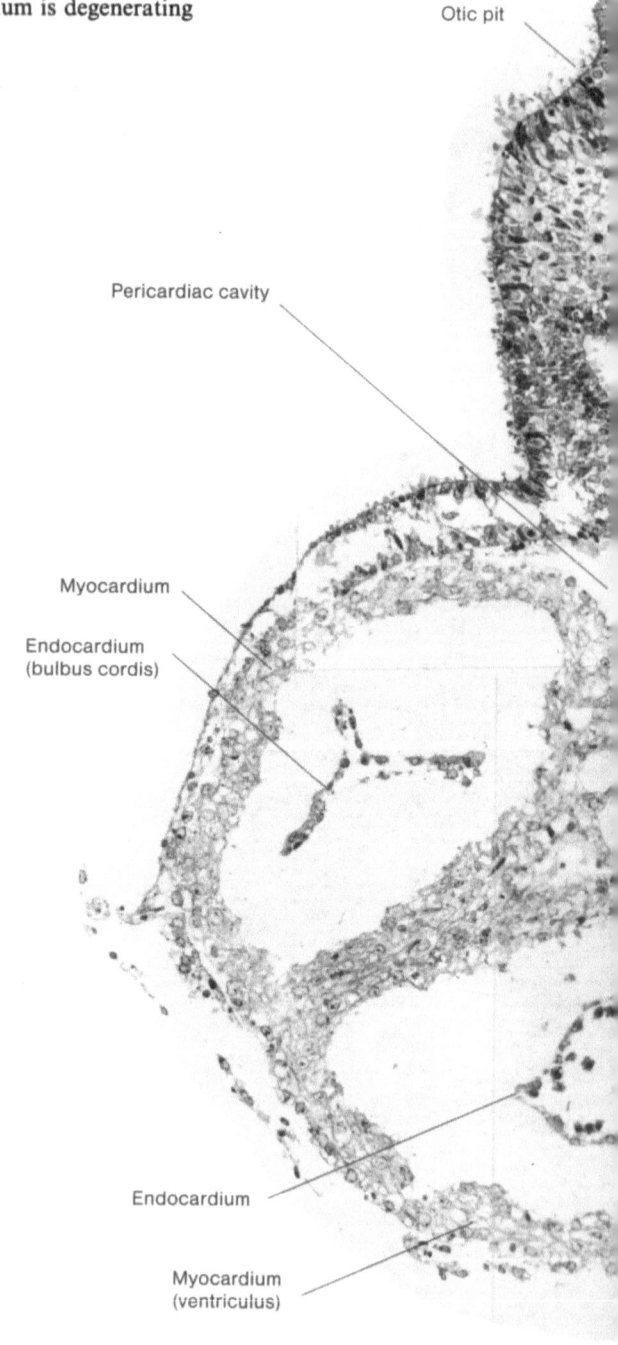

Otic pit

Pericardiac cavity

Myocardium

Endocardium
(bulbus cordis)

Endocardium

Myocardium
(ventriculus)

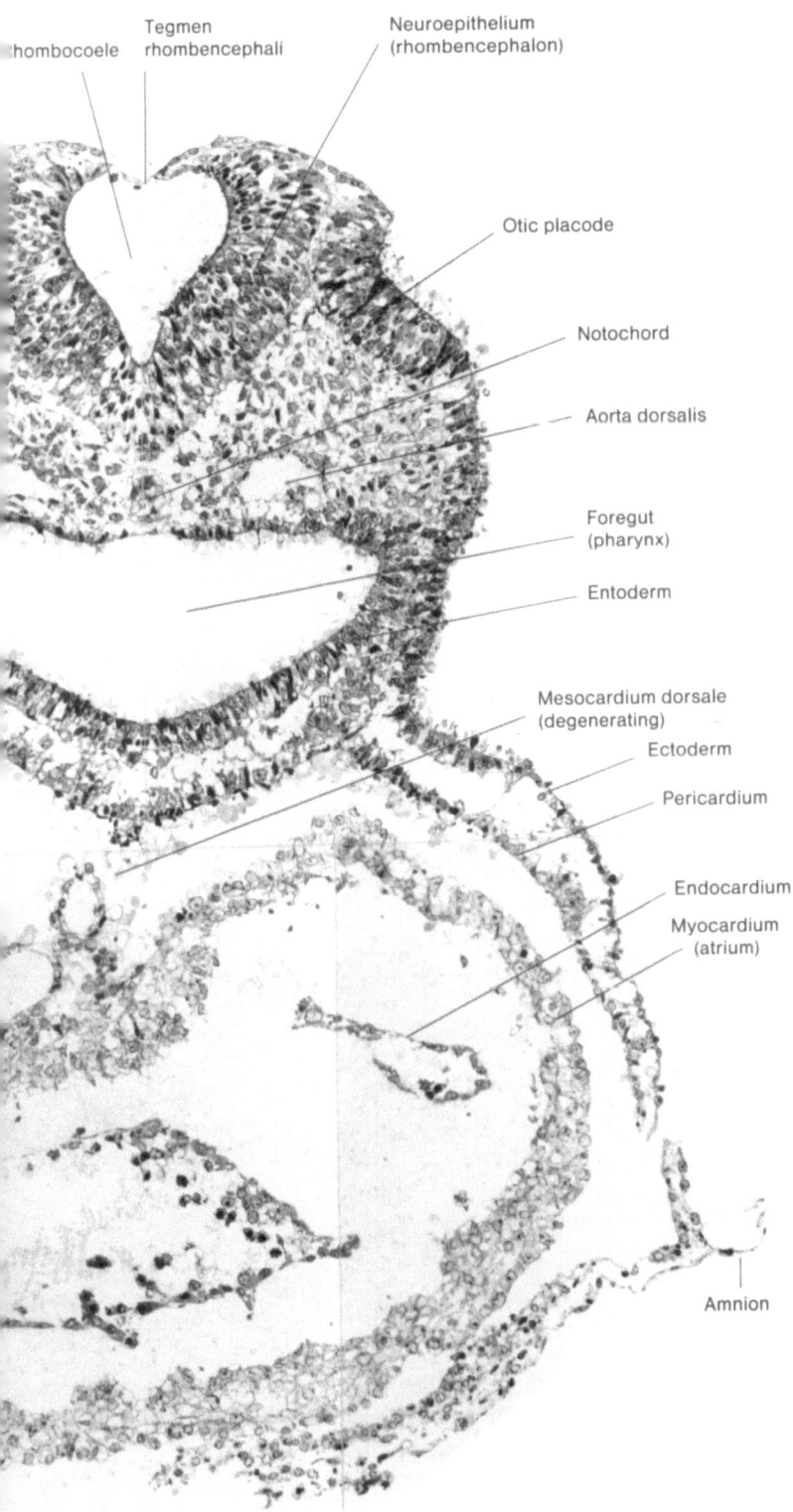

Rhombocoele

Tegmen
rhombencephali

Neuroepithelium
(rhombencephalon)

Otic placode

Notochord

Aorta dorsalis

Foregut
(pharynx)

Entoderm

Mesocardium dorsale
(degenerating)

Ectoderm

Pericardium

Endocardium

Myocardium
(atrium)

Amnion

Fig. 22. Section through second branchial groove and second branchial pouch (13–1). The lumen of the foregut, the pharynx, widens transversely to form the second branchial pouch, which corresponds to a faint depression of the body surface, the second branchial groove. Lateral to the dorsal portion of the rhombencephalon the rest of the otic placode is seen. A faint lateral bulge between the second branchial groove and the otic placode represents the third branchial arch. The myocardial tube elongates transversely and contains the endocardial lumina of the bulbus cordis, ventriculus, and atrium. The dorsal mesocardium is here also degenerating. The pericardiac cavity is first closed by a thin mesenchymal cell layer, indicating the anterior end of the septum transversum

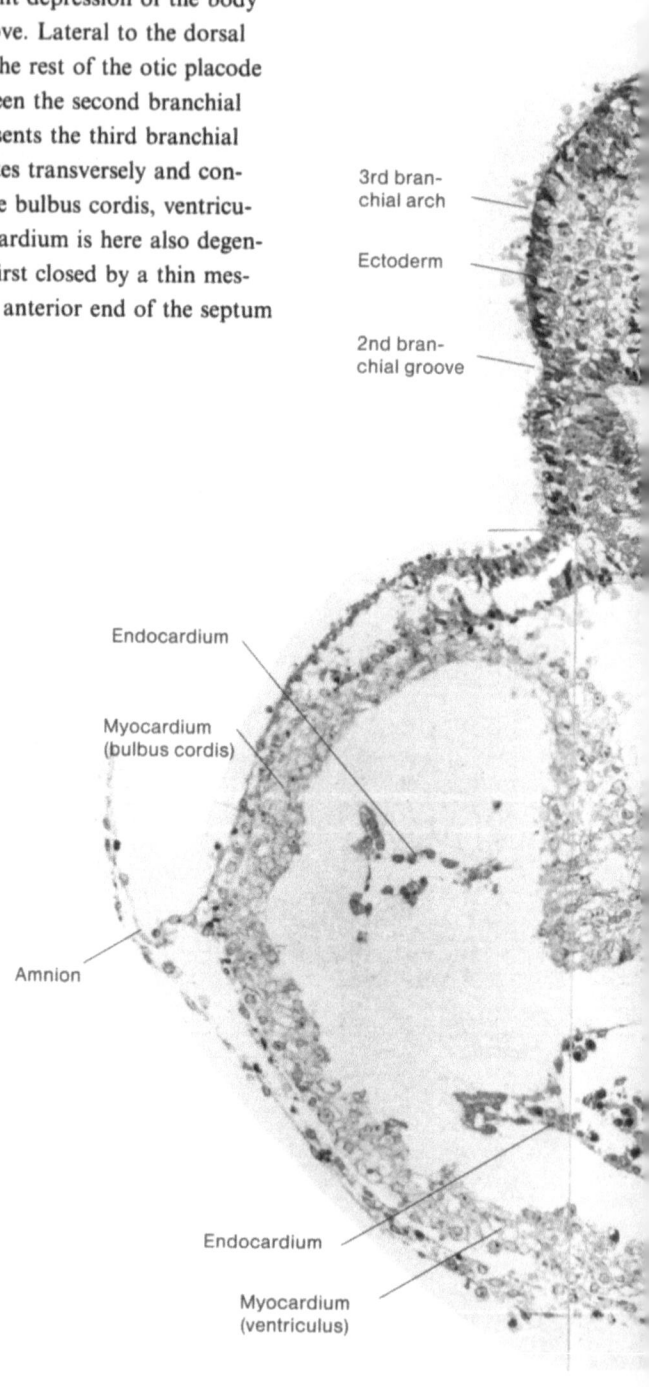

3rd branchial arch

Ectoderm

2nd branchial groove

Endocardium

Myocardium (bulbus cordis)

Amnion

Endocardium

Myocardium (ventriculus)

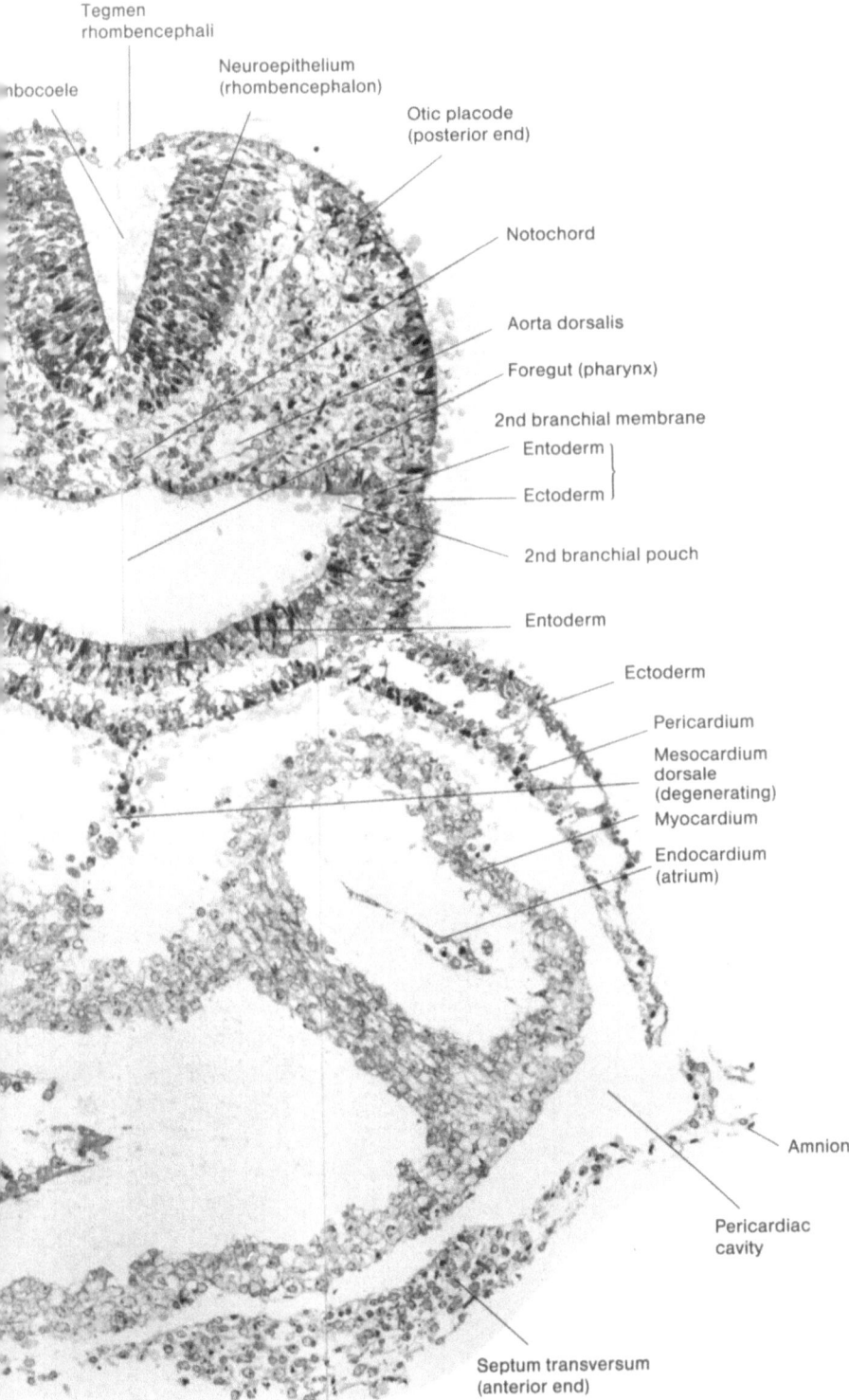

Tegmen
rhombencephali

Neuroepithelium
(rhombencephalon)

nbocoele

Otic placode
(posterior end)

Notochord

Aorta dorsalis

Foregut (pharynx)

2nd branchial membrane

Entoderm

Ectoderm

2nd branchial pouch

Entoderm

Ectoderm

Pericardium

Mesocardium
dorsale
(degenerating)

Myocardium

Endocardium
(atrium)

Amnion

Pericardiac
cavity

Septum transversum
(anterior end)

Fig. 23. Section through anterior portion of septum transversum (17–1). The rhombencephalon now shows a characteristic triangular configuration with ventrally oriented apex and dorsally situated base. The latter consists of a very thin two-cell-layer membrane, the tegmen rhombencephali, which has here its greatest width of about 130 μm. The notochord is separated from both the ventral aspect of the rhombencephalon and the dorsal wall of the foregut. The myocardial tube of the ventriculus is still large, but the endocardial tube is no longer seen. The atrium elongates to the right, dorsal to the ventriculus, and its left and right halves each contain one endocardiac lumen. The dorsal mesocardium does not exist here

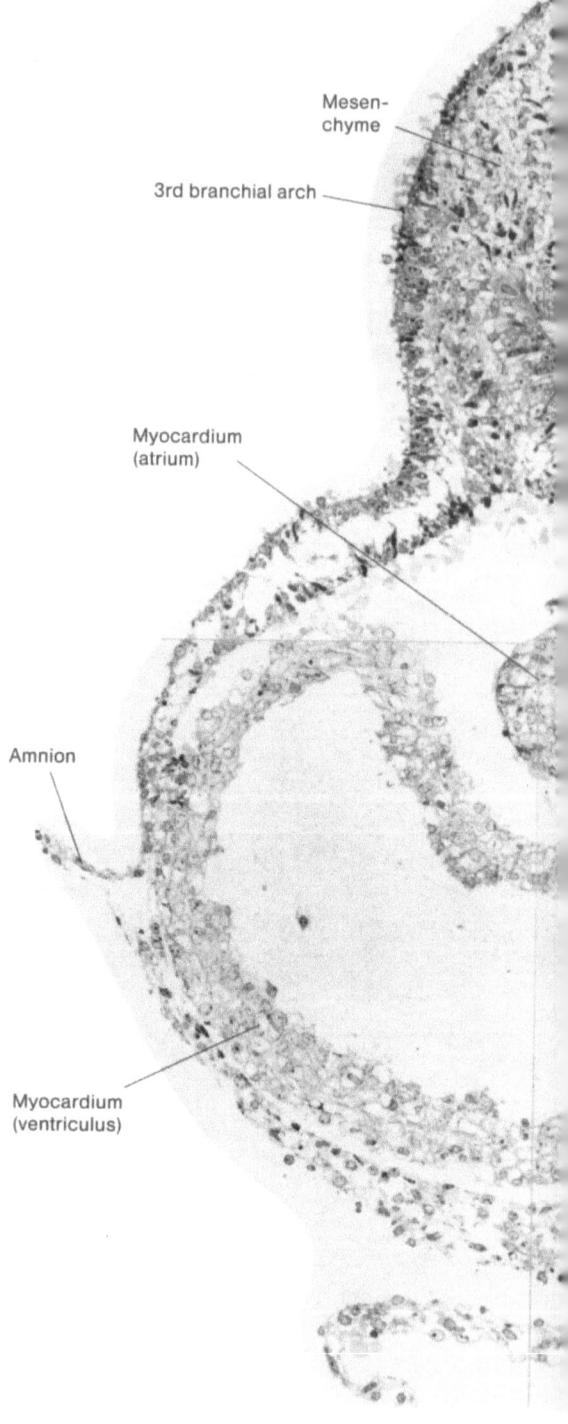

Mesen-chyme

3rd branchial arch

Myocardium (atrium)

Amnion

Myocardium (ventriculus)

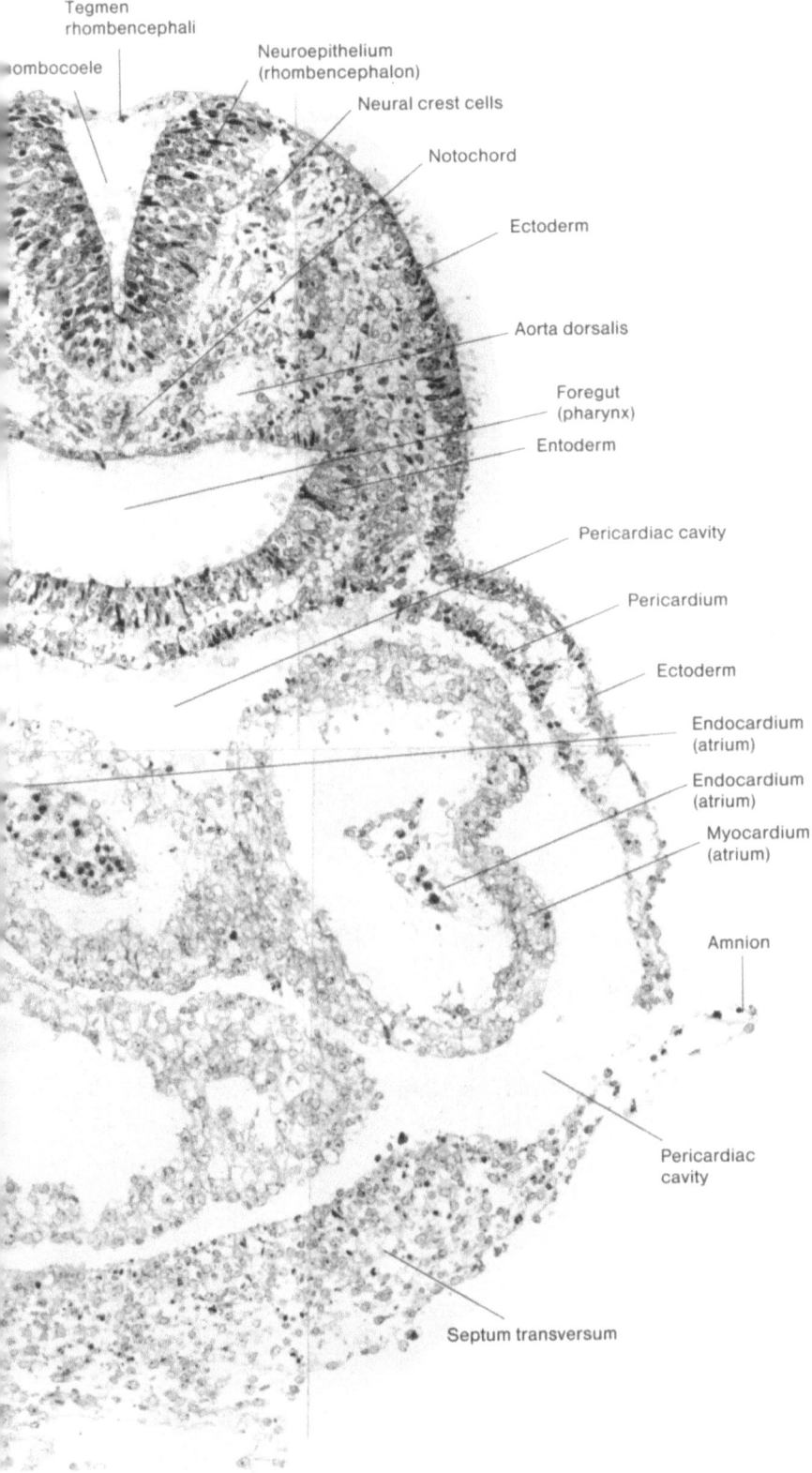

Tegmen
rhombencephali

Neuroepithelium
(rhombencephalon)

rhombocoele

Neural crest cells

Notochord

Ectoderm

Aorta dorsalis

Foregut
(pharynx)

Entoderm

Pericardiac cavity

Pericardium

Ectoderm

Endocardium
(atrium)

Endocardium
(atrium)

Myocardium
(atrium)

Amnion

Pericardiac
cavity

Septum transversum

Fig. 24. Section through posterior end of ventriculus (20). From the dorsolateral extremity of the rhombencephalon neural crest cells are migrating as small cell groups. The atrium enlarges transversely to the right and contains an elongated endocardial tube. At the mid-dorsal wall of the pericardiac cavity a vestige of the dorsal mesocardium is seen. The septum transversum distinctly increases in size

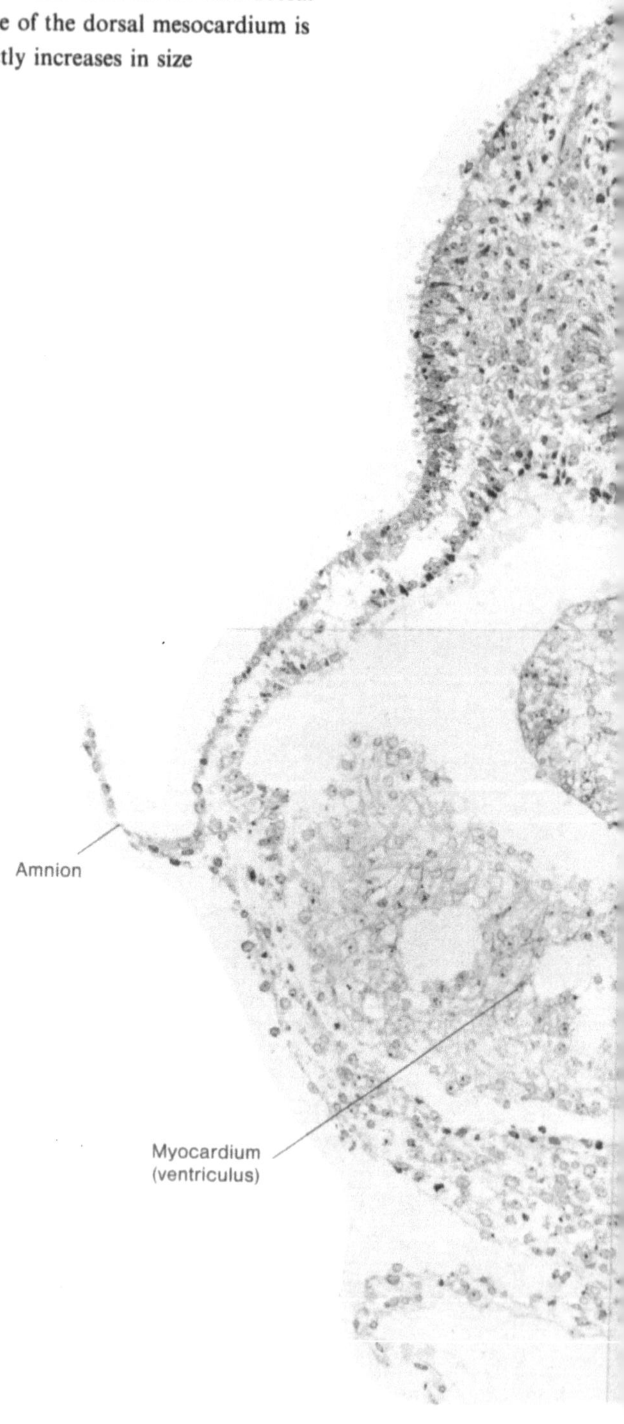

Amnion

Myocardium
(ventriculus)

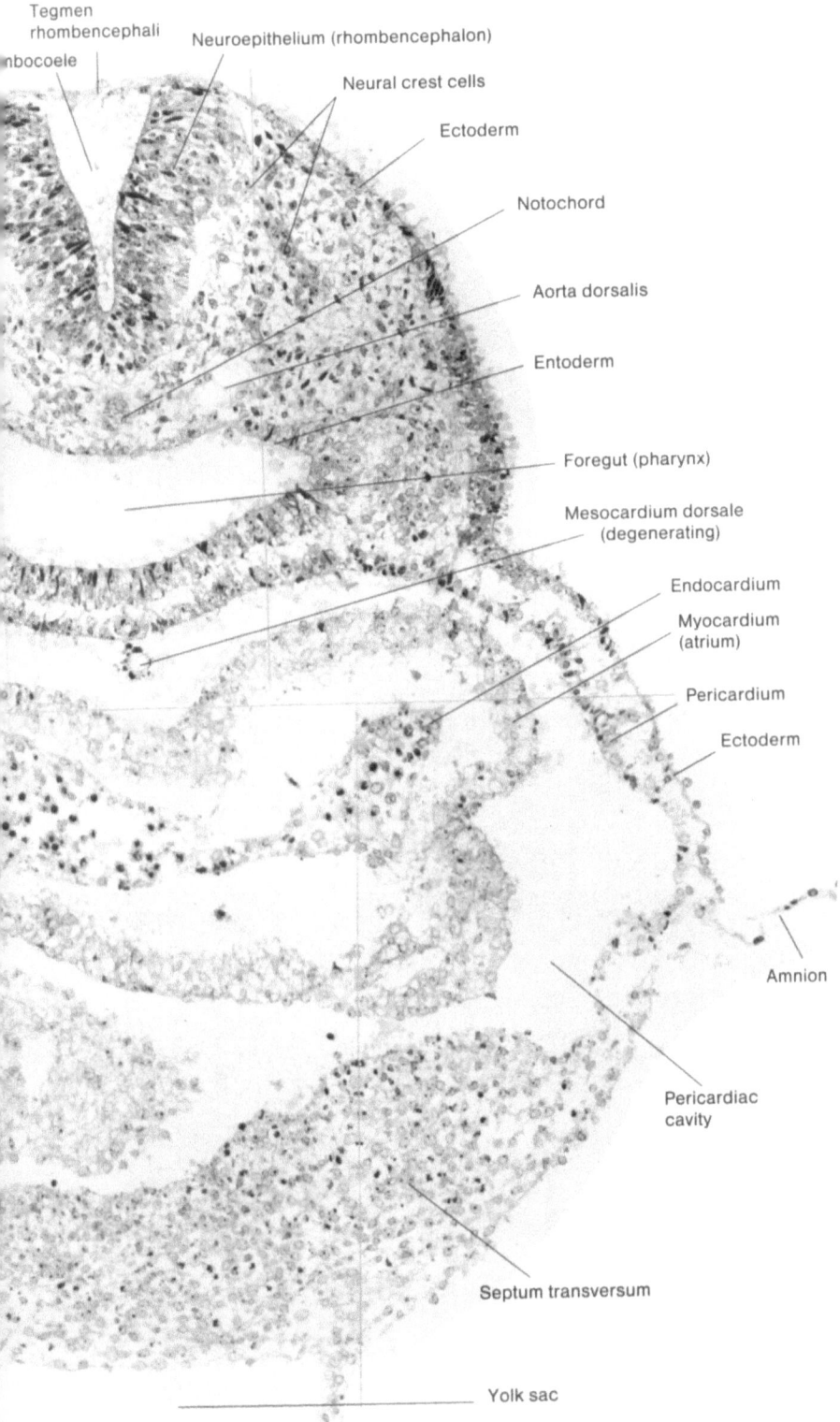

Tegmen
rhombencephali

Neuroepithelium (rhombencephalon)

nbocoele

Neural crest cells

Ectoderm

Notochord

Aorta dorsalis

Entoderm

Foregut (pharynx)

Mesocardium dorsale
(degenerating)

Endocardium

Myocardium
(atrium)

Pericardium

Ectoderm

Amnion

Pericardiac
cavity

Septum transversum

Yolk sac

Fig. 25. Section through posterior portion of atrium (25–1). The tegmen rhombencephali diminishes in width so that the rhombocoele becomes narrower. In the pericardiac cavity only the transversely situated atrium is seen. The endocardiac tube slightly enlarges at its left and right extremities. The septum transversum appears as a massive mesenchymal cell aggregation separating the pericardiac cavity from that of the yolk sac

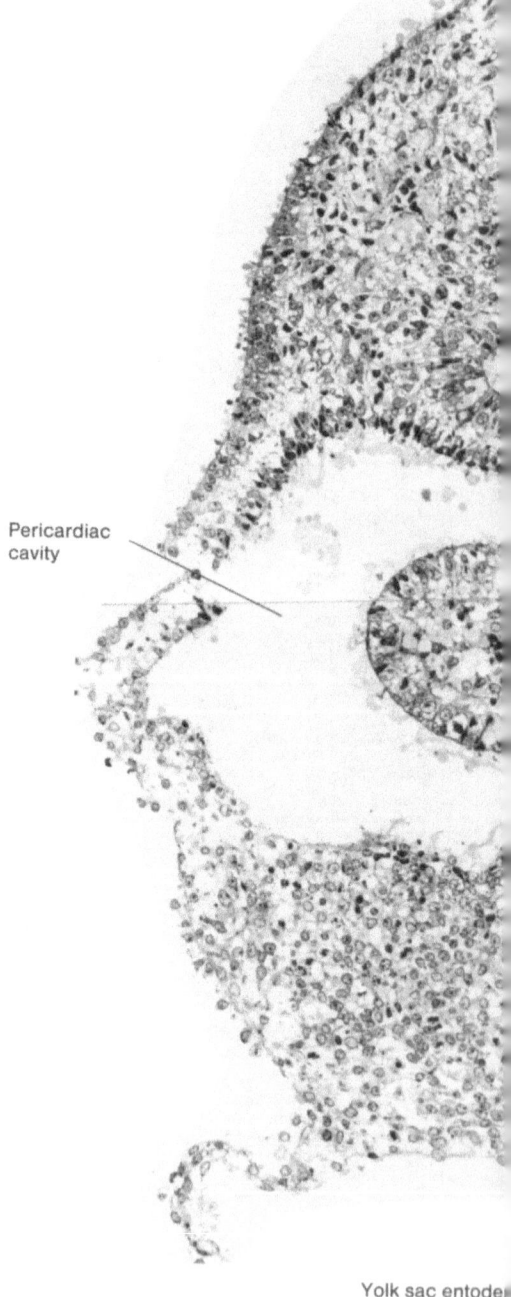

Pericardiac cavity

Yolk sac entoderm

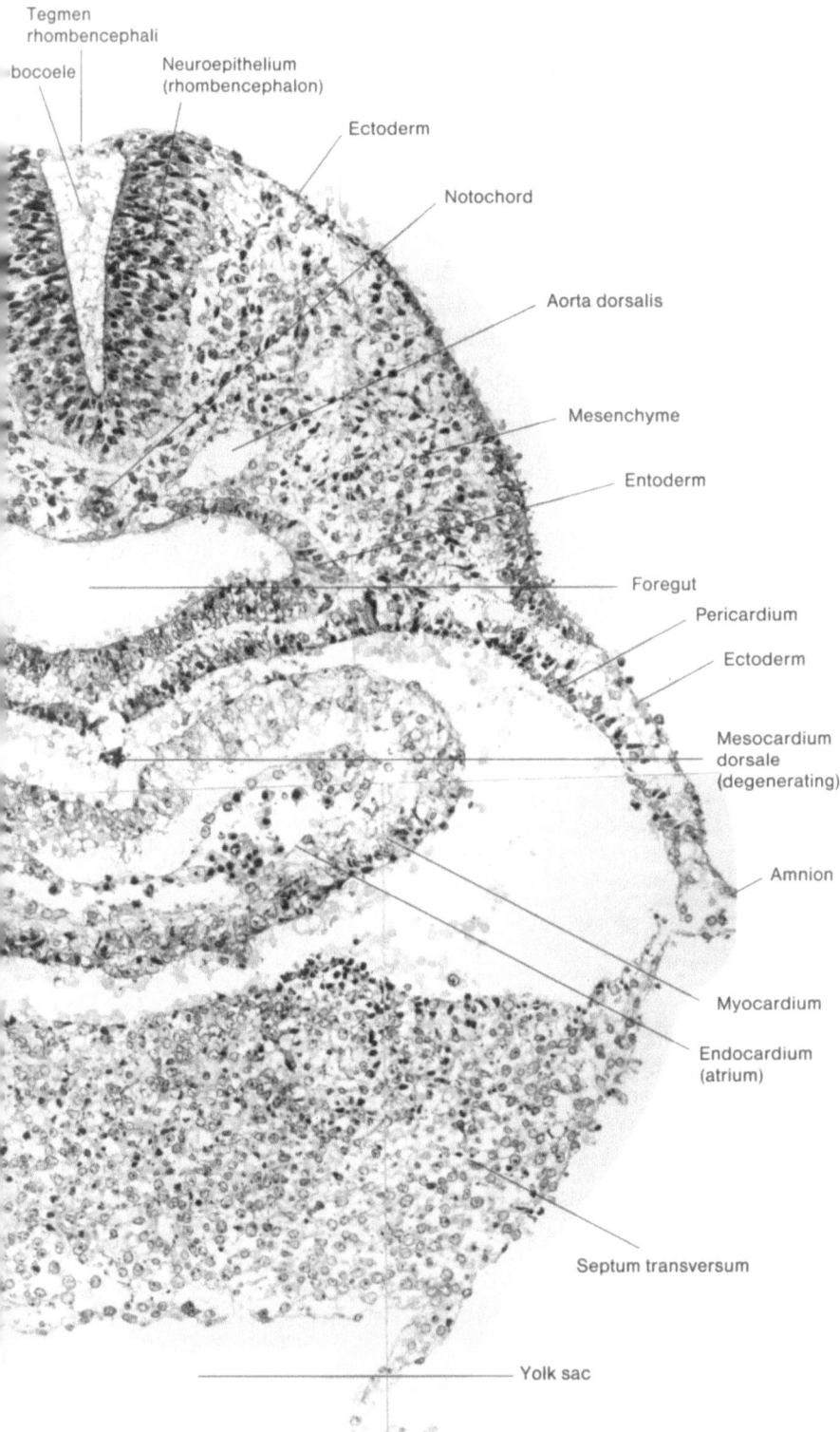

Tegmen
rhombencephali

bocoele

Neuroepithelium
(rhombencephalon)

Ectoderm

Notochord

Aorta dorsalis

Mesenchyme

Entoderm

Foregut

Pericardium

Ectoderm

Mesocardium
dorsale
(degenerating)

Amnion

Myocardium

Endocardium
(atrium)

Septum transversum

Yolk sac

43

Fig. 26. Section through first somite and hepatic primordium (29–1). Lateral to the dorsal portion of the rhombencephalon a small cell group showing the columnar epithelial cell arrangement is seen on each side. This represents the dermomyotome of the first somite, whose sclerotome shows no more epithelial cell arrangement but has already migrated out to become mesenchyme. The atrium diminishes in size and its endocardial lumen divides into two indicating the left and right sinus venosi. The hepatic primordium appears as a large and conspicuous cell mass at the mid-ventral portion of the septum transversum

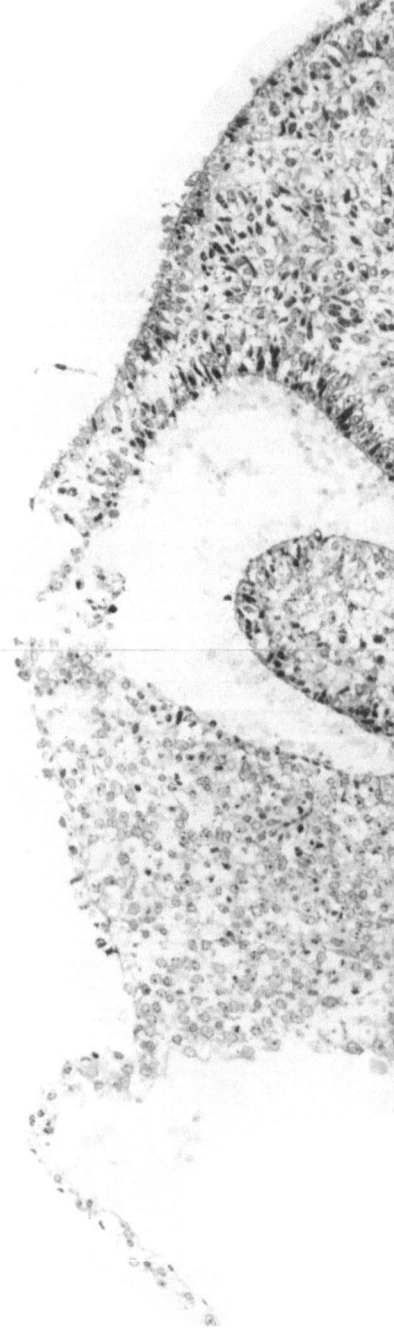

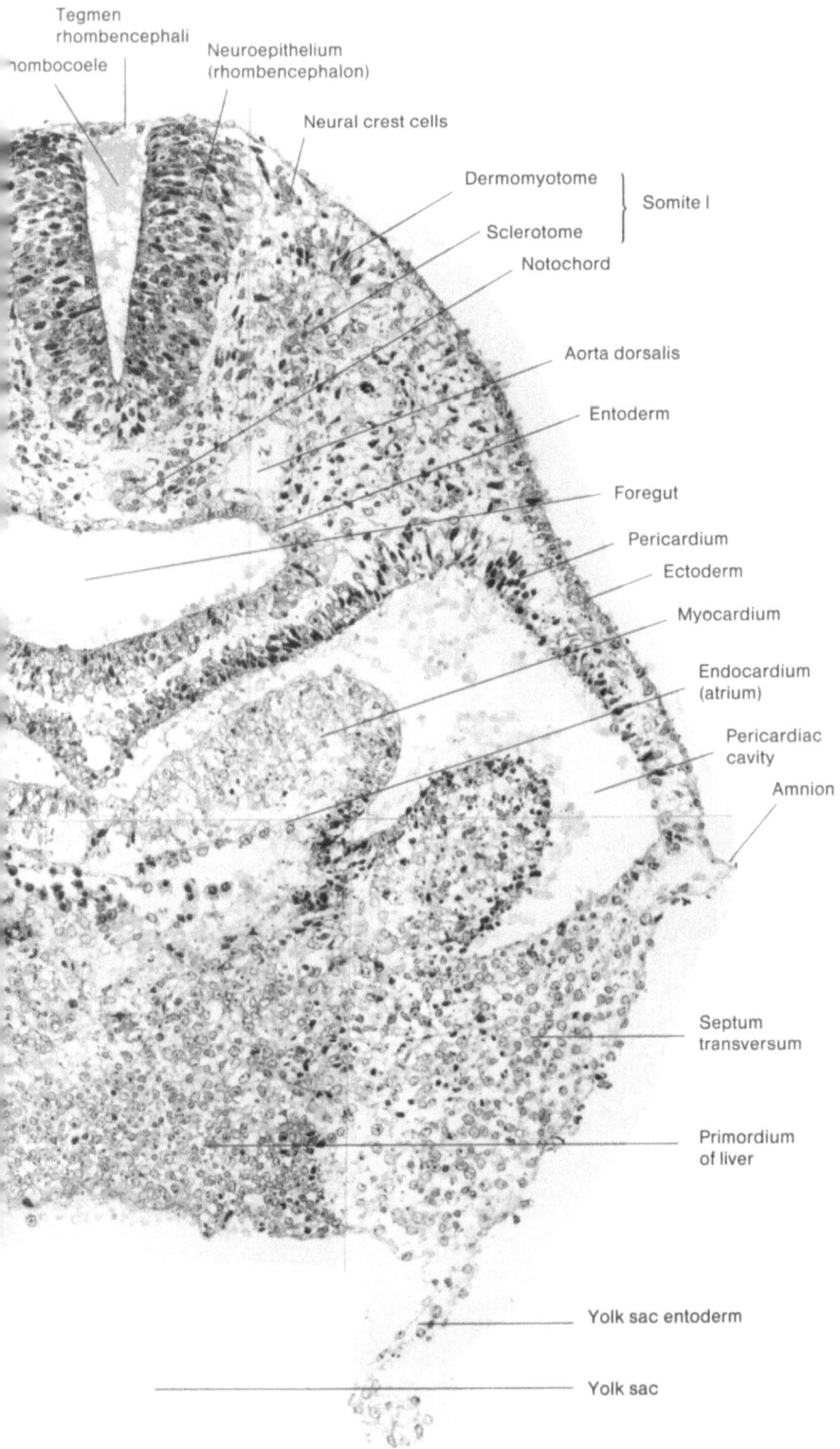

Tegmen rhombencephali

rhombocoele

Neuroepithelium (rhombencephalon)

Neural crest cells

Dermomyotome

Sclerotome

} Somite I

Notochord

Aorta dorsalis

Entoderm

Foregut

Pericardium

Ectoderm

Myocardium

Endocardium (atrium)

Pericardiac cavity

Amnion

Septum transversum

Primordium of liver

Yolk sac entoderm

Yolk sac

Fig. 27. Section through posterior portion of first somite and posterior end of foregut (32–2). The dorsal wall of the rhombocoele consists no longer of the tegmen rhombence-phali but of a slightly thicker membrane, the roof plate. The entodermal epithelium lining of the ventral wall of the foregut elongates ventrally and connects with the entodermal lining of the yolk sac. Lateral to the foregut is located the coelomic duct, and ventral to it the sinus venosus is embedded in the mesenchymal cells of the septum transversum. On the outer surface of the yolk sac blood islets are evident

Ecto-
derm

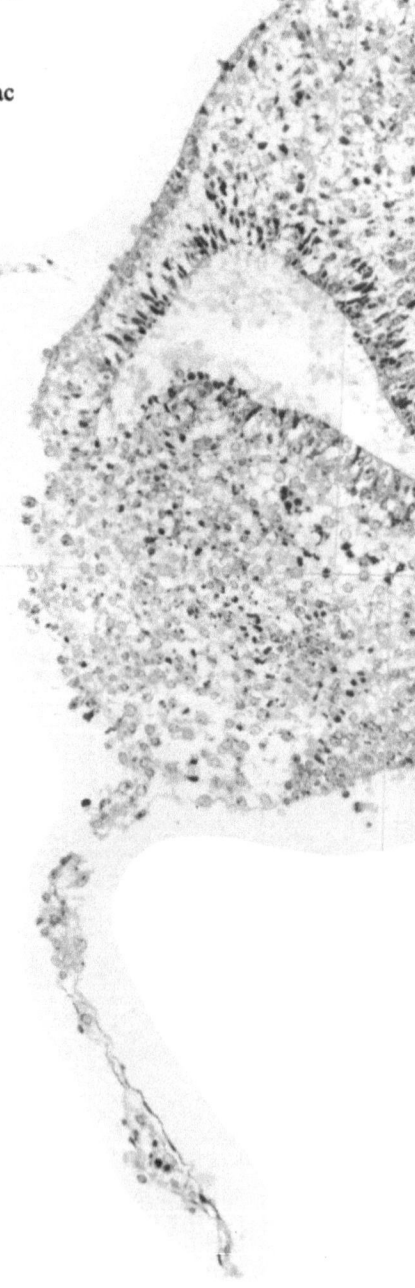

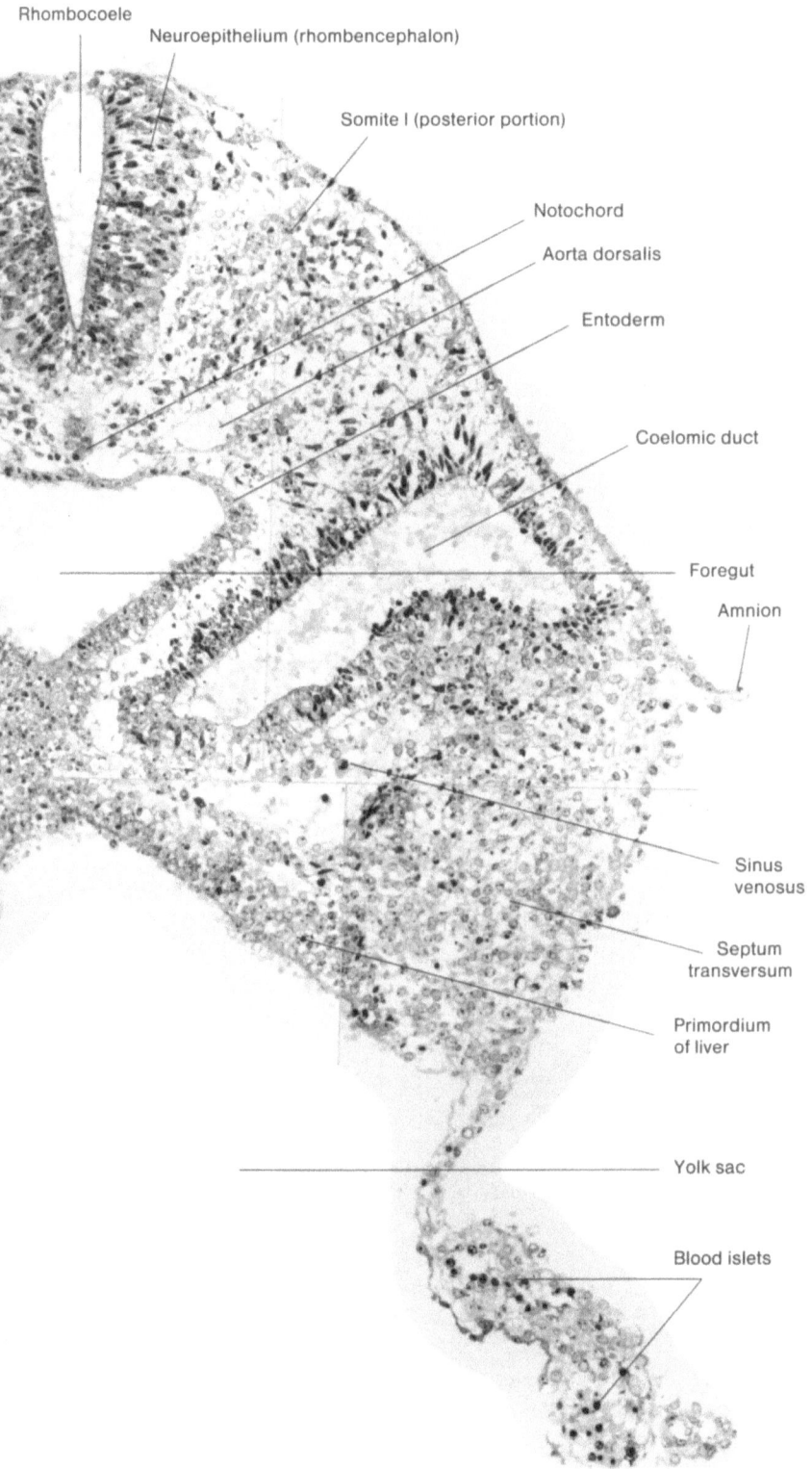

Rhombocoele

Neuroepithelium (rhombencephalon)

Somite I (posterior portion)

Notochord

Aorta dorsalis

Entoderm

Coelomic duct

Foregut

Amnion

Sinus venosus

Septum transversum

Primordium of liver

Yolk sac

Blood islets

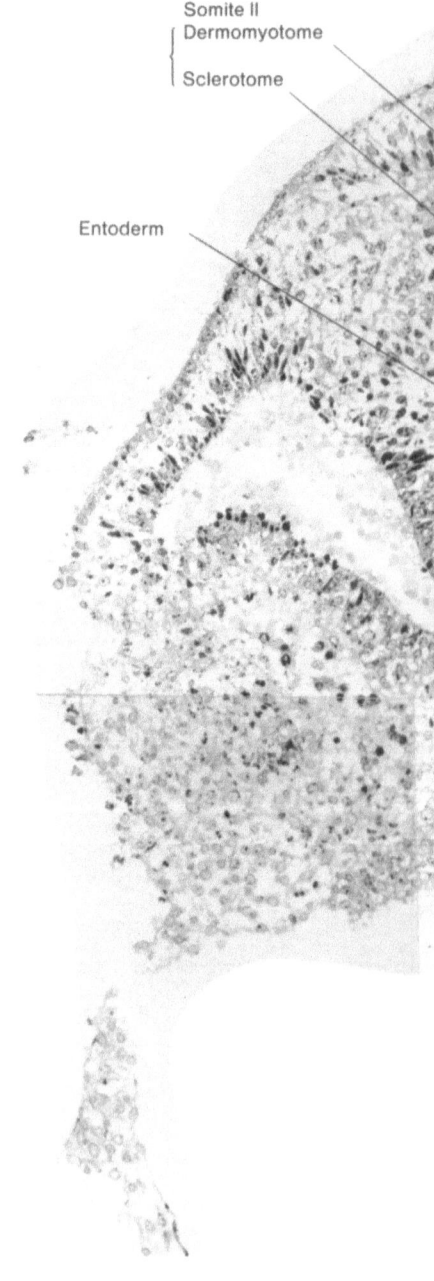

Somite II
Dermomyotome

Sclerotome

Entoderm

Fig. 28. Section through second somite and anterior intestinal portal (33–3). The lumen of the foregut opens ventrally into the yolk sac cavity; thus, the foregut passes on to the midgut. Ventral to this passage the inner surface of the yolk sac has a conspicuous cell

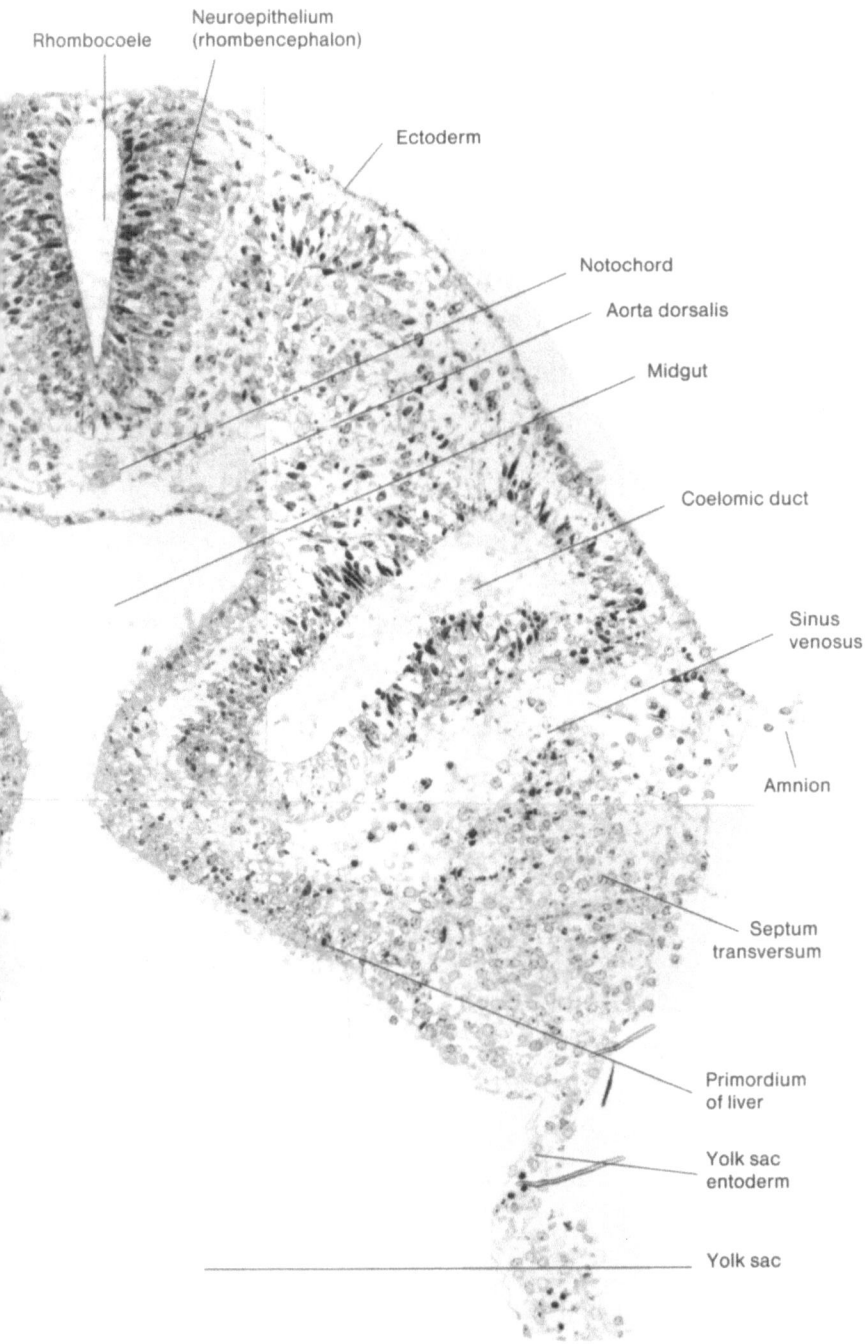

Rhombocoele

Neuroepithelium
(rhombencephalon)

Ectoderm

Notochord

Aorta dorsalis

Midgut

Coelomic duct

Sinus
venosus

Amnion

Septum
transversum

Primordium
of liver

Yolk sac
entoderm

Yolk sac

condensation representing the hepatic primordium, and lateral to it there are the coelomic duct and sinus venosus on each side. The second somite, like the first, consists of dorsolateral dermomyotome and ventromedial sclerotome of mesenchymal cell arrangement

49

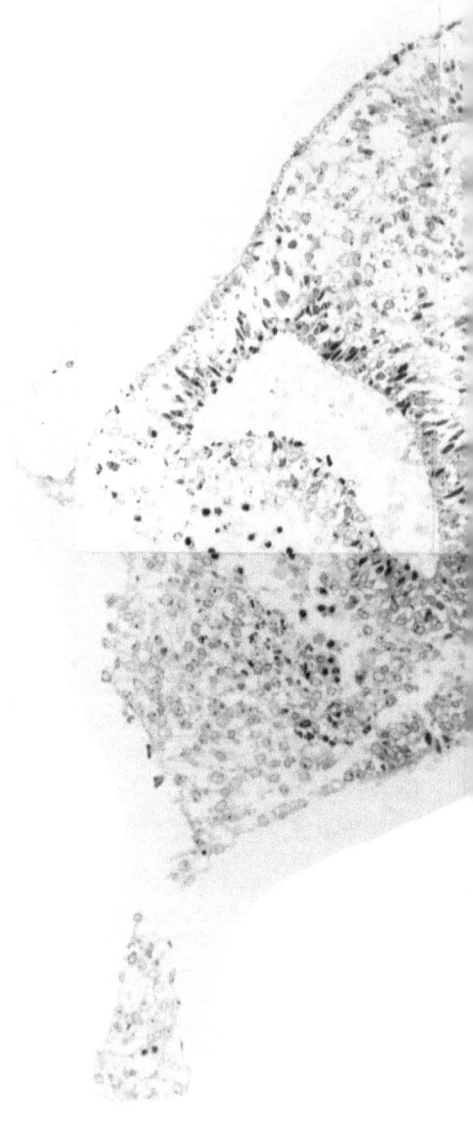

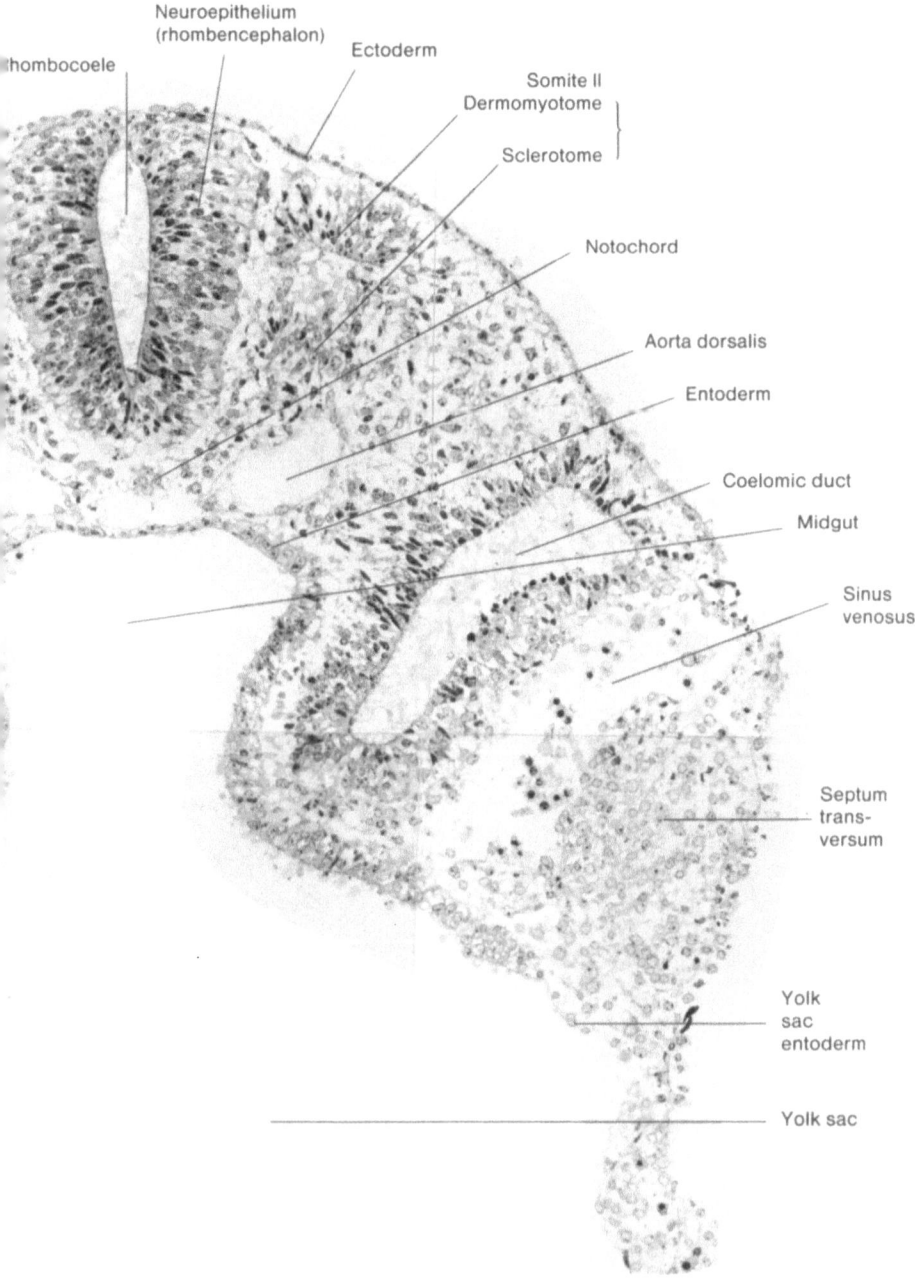

Rhombocoele

Neuroepithelium
(rhombencephalon)

Ectoderm

Somite II
Dermomyotome

Sclerotome

Notochord

Aorta dorsalis

Entoderm

Coelomic duct

Midgut

Sinus
venosus

Septum
trans-
versum

Yolk
sac
entoderm

Yolk sac

Fig. 29. Section through posterior portion of second somite (35–2). Except for the transverse enlargement of the communication between the midgut and the yolk sac, the morphological features of this section are similar to those seen in Fig. 28

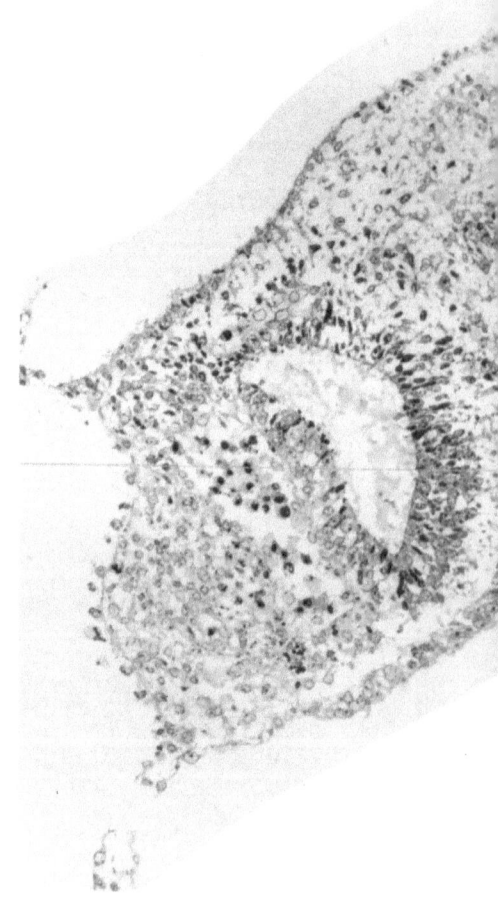

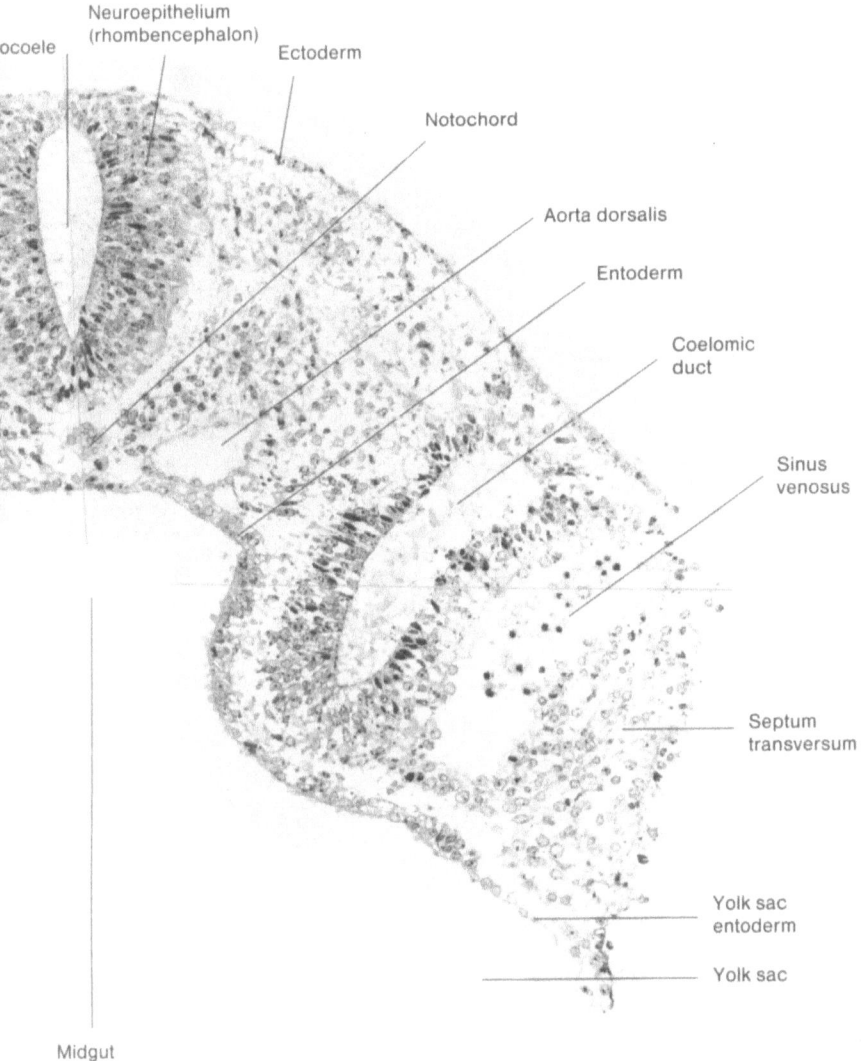

Neuroepithelium
(rhombencephalon)

ocoele

Ectoderm

Notochord

Aorta dorsalis

Entoderm

Coelomic
duct

Sinus
venosus

Septum
transversum

Yolk sac
entoderm

Yolk sac

Midgut

Fig. 30. Section through interspace between second and third somites (36–3). Lateral to the neural tube, the rhombencephalon, this interspace is loosely filled with mesenchymal cells. As the transverse widening of the communication between the midgut and the yolk sac progresses, the embryonic body itself attains a transverse configuration; for example, the sinus venosus is now located not ventral but lateral to the coelomic duct

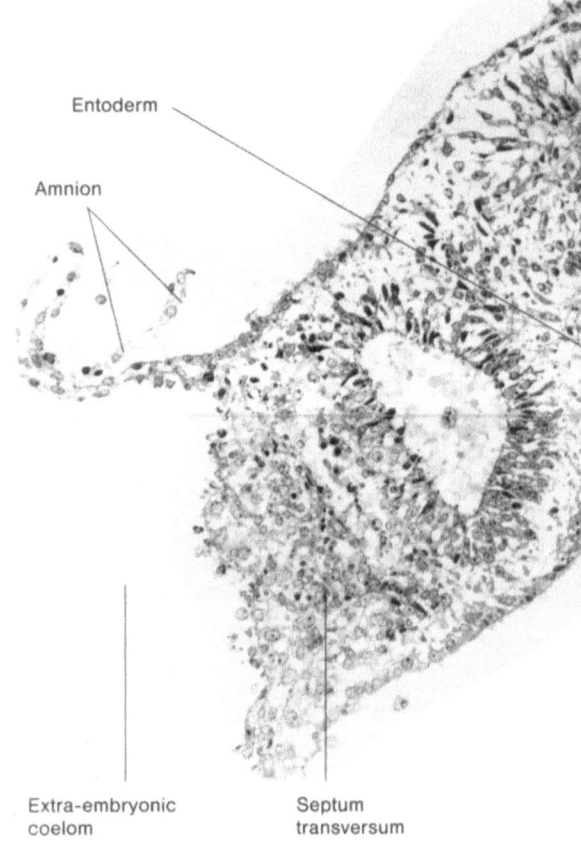

Entoderm

Amnion

Extra-embryonic
coelom

Septum
transversum

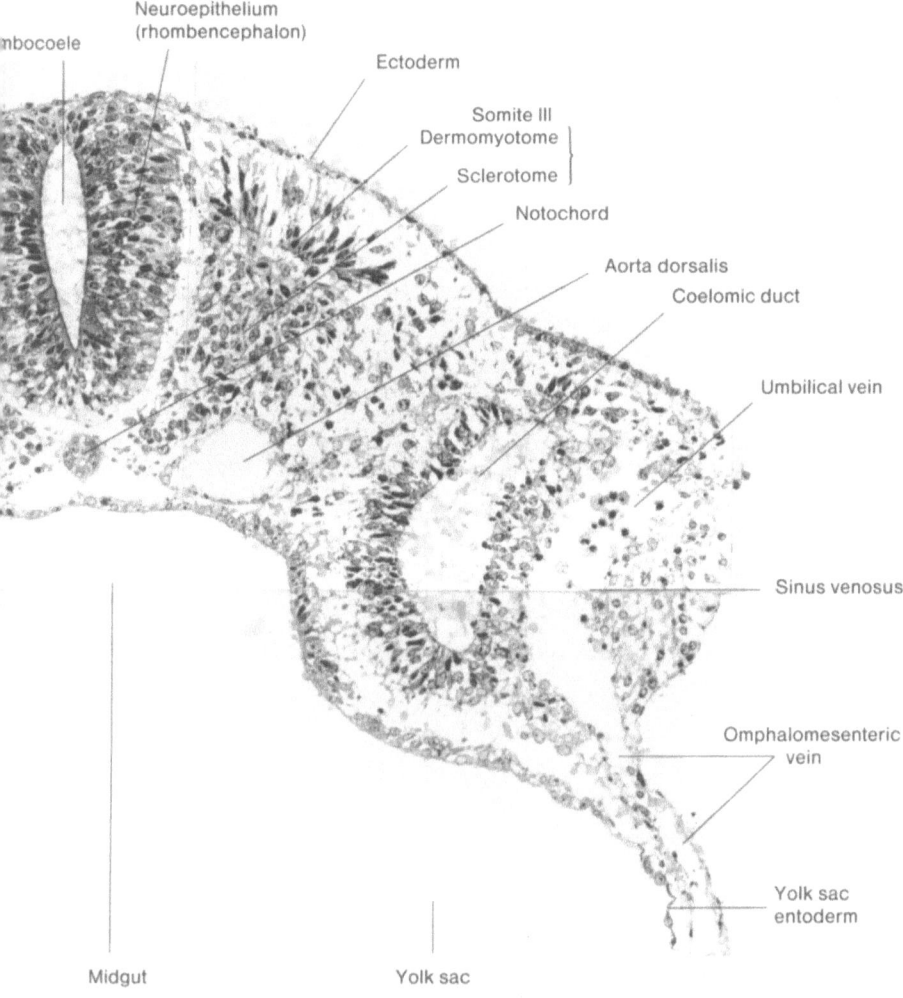

Neuroepithelium
(rhombencephalon)

nbocoele

Ectoderm

Somite III
Dermomyotome

Sclerotome

Notochord

Aorta dorsalis

Coelomic duct

Umbilical vein

Sinus venosus

Omphalomesenteric
vein

Yolk sac
entoderm

Midgut

Yolk sac

Fig. 31. Section through third somite (39). The third somite is obviously larger and shows a more densely packed cell arrangement than the first two. The notochord, with a distinct round contour, is located on the median line between the neural tube and dorsal wall of the midgut, separated from both of these structures. On the left side the sinus venosus accepts an omphalomesenteric vein and the umbilical vein at its ventral and dorsolateral extremities respectively. The septum transversum still remains as a mesenchymal tissue lateral to the sinus venosus and faces, laterally, the extra-embryonic coelom

Somite IV
Dermomyotome

Sclerotome

Amnion

Septum
transversum

entral canal

Neuroepithelium
(spinal cord)

Ectoderm

Notochord

Aorta dorsalis

Entoderm

Coelomic duct

Umbilical
vein

Extra-
embryonic
coelom

Yolk sac
entoderm

Midgut

Yolk sac

Fig. 32. Section through fourth somite (44–3). The neural tube now represents the spinal cord and follows the rhombencephalon without any boundary. The umbilical vein is located at the dorsolateral corner of the body, where the surface ectoderm passes suddenly on to the amniotic epithelium. The morphological features of this section are in general quite similar to those shown in Fig. 31

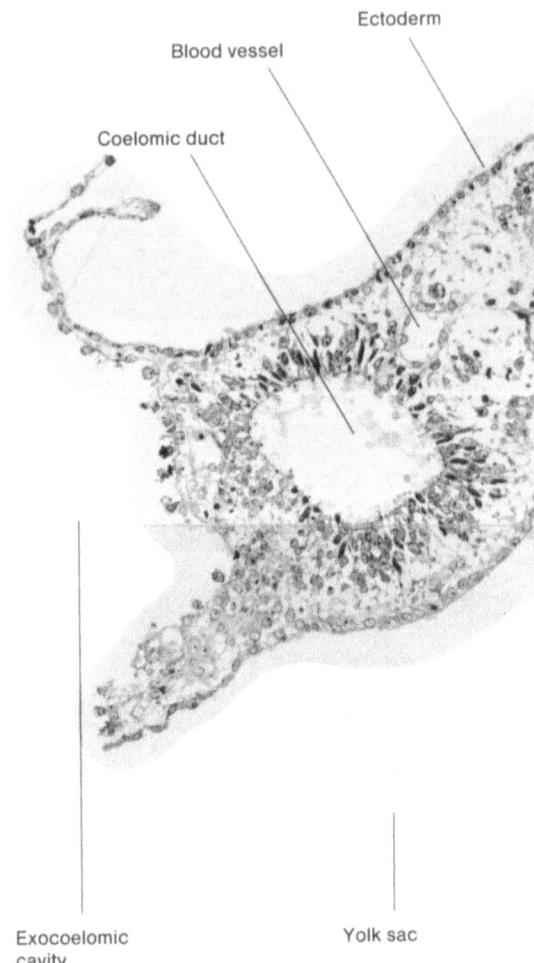

Ectoderm

Blood vessel

Coelomic duct

Exocoelomic
cavity

Yolk sac

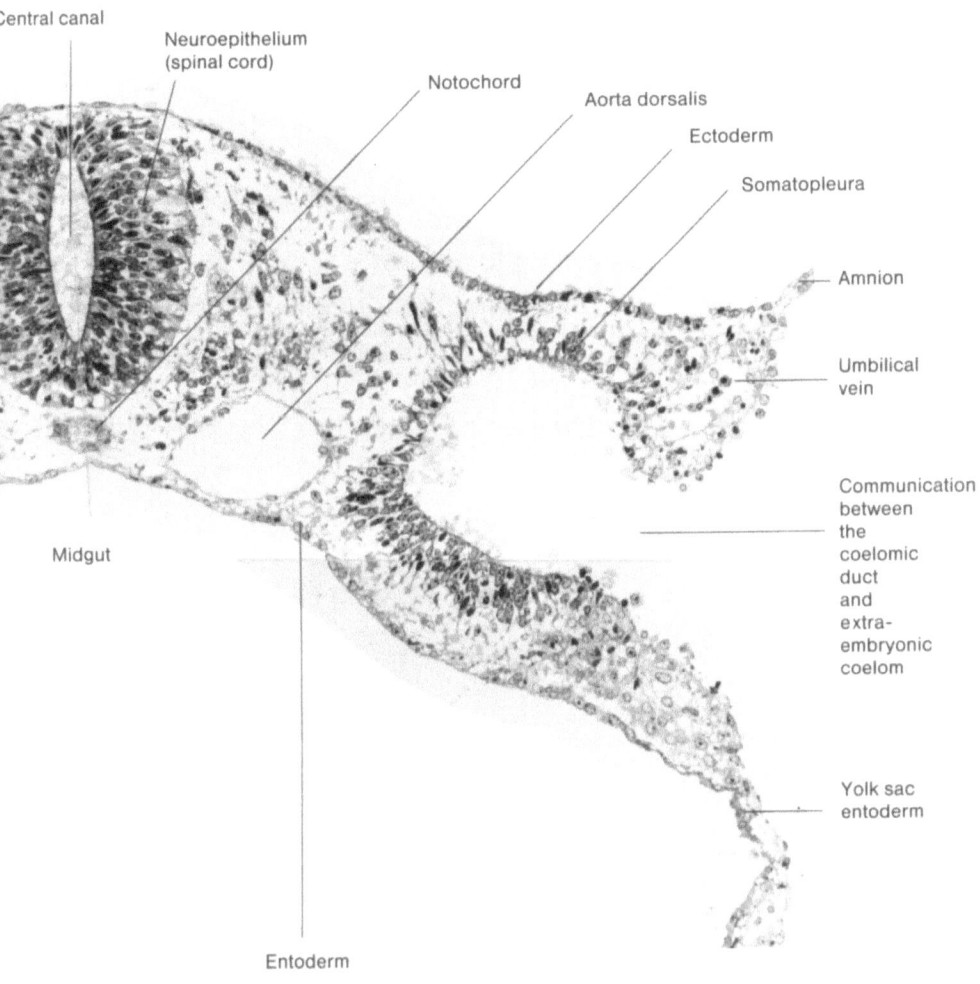

Central canal

Neuroepithelium
(spinal cord)

Notochord

Aorta dorsalis

Ectoderm

Somatopleura

Amnion

Umbilical
vein

Communication
between
the
coelomic
duct
and
extra-
embryonic
coelom

Yolk sac
entoderm

Midgut

Entoderm

Fig. 33. Section through interspace between fourth and fifth somites (50–2). In the inter-space mesenchymal cells are loosely scattered. On the left side the coelomic duct opens laterally into the extra-embryonic coelom

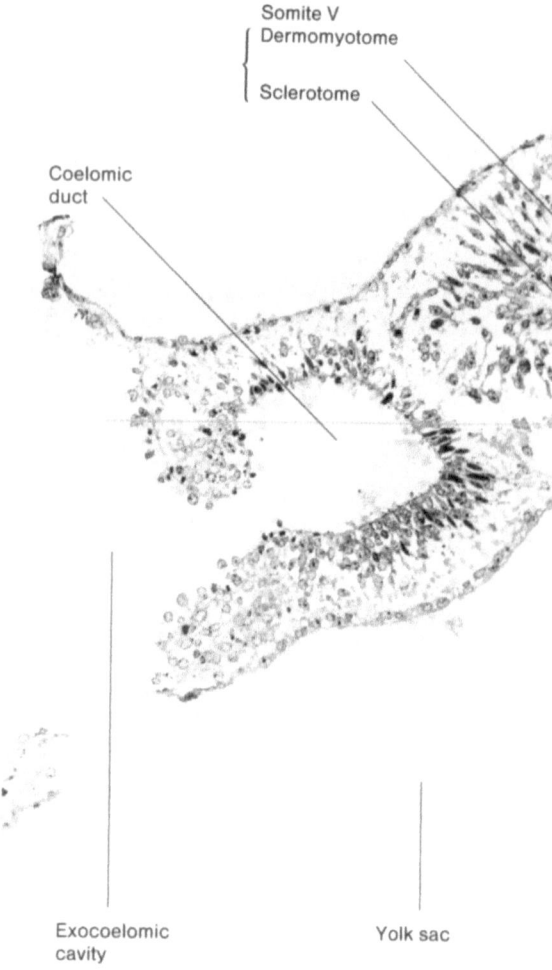

Somite V
Dermomyotome

Sclerotome

Coelomic
duct

Exocoelomic
cavity

Yolk sac

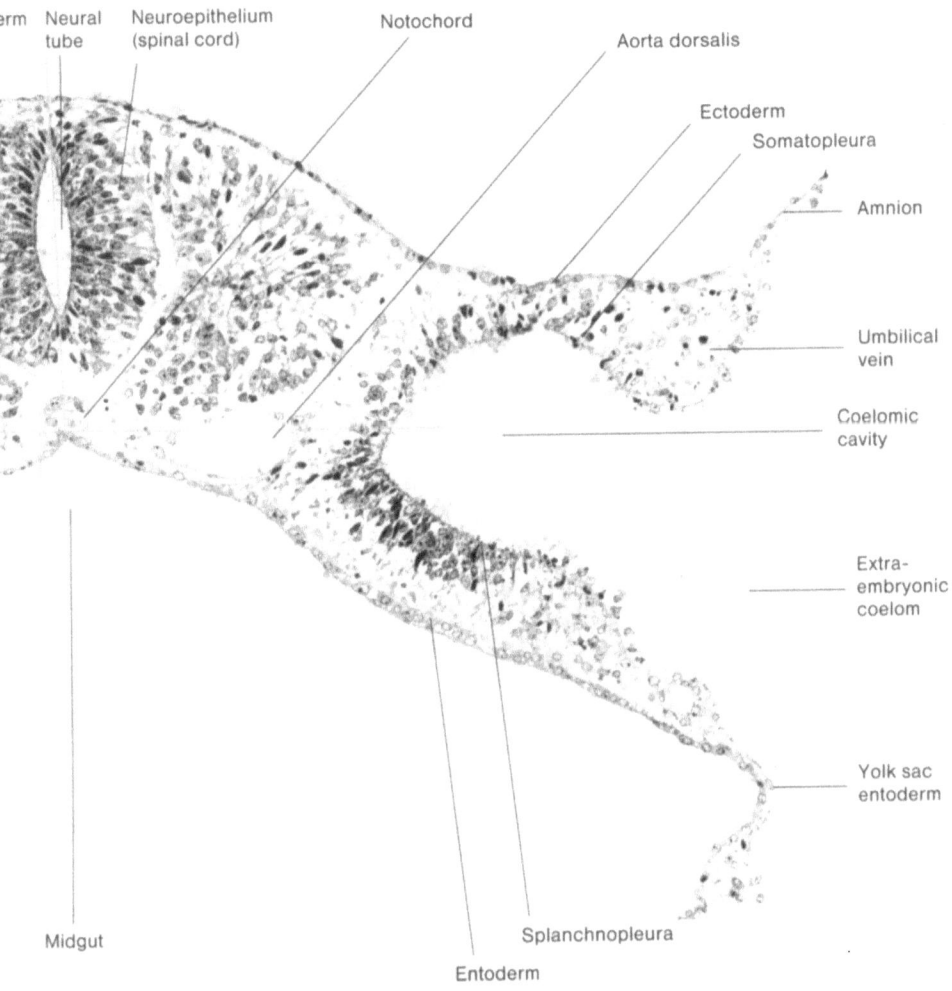

erm Neural | Neuroepithelium | Notochord | Aorta dorsalis
tube | (spinal cord)

Ectoderm
Somatopleura

Amnion

Umbilical
vein

Coelomic
cavity

Extra-
embryonic
coelom

Yolk sac
entoderm

Midgut | Splanchnopleura

Entoderm

Fig. 34. Section through fifth somite (53–2). The coelomic duct opens laterally also on the right side into the extra-embryonic coelom. The epithelium lining the coelomic cavity thus divides into two: the splanchnopleura underlining the entoderm and the somatopleura underlining the ectoderm. The umbilical vein runs at the lateral extremity of the latter. The ventral aspect of the notochord is again in contact with the mid-dorsal wall of the midgut

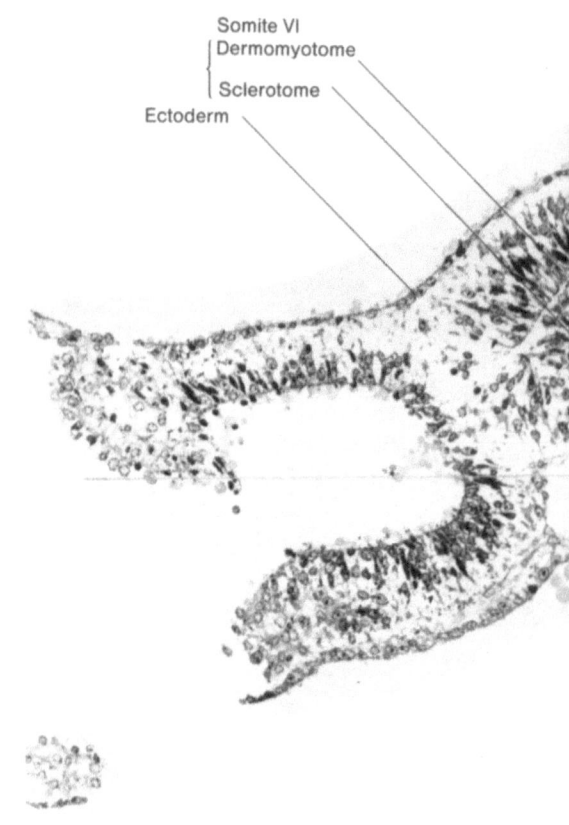

Somite VI
Dermomyotome

Sclerotome

Ectoderm

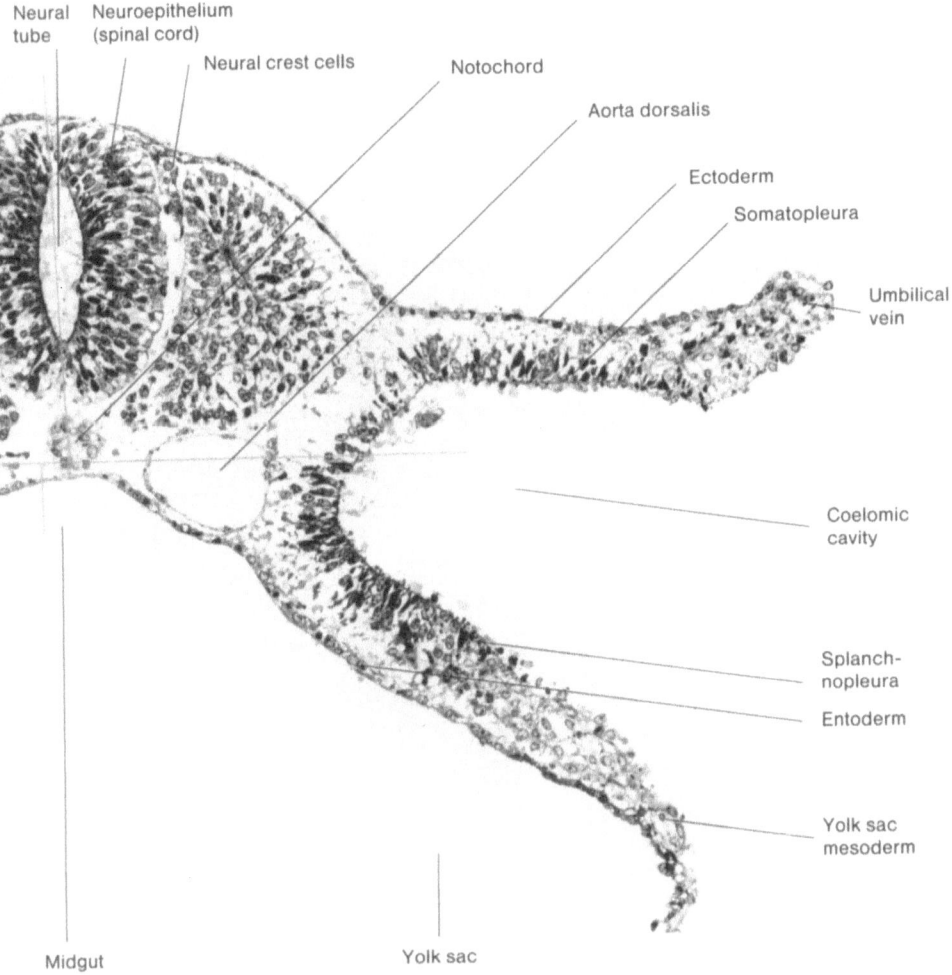

Neural tube

Neuroepithelium (spinal cord)

Neural crest cells

Notochord

Aorta dorsalis

Ectoderm

Somatopleura

Umbilical vein

Coelomic cavity

Splanch- nopleura

Entoderm

Yolk sac mesoderm

Midgut

Yolk sac

Fig. 35. Section through sixth somite (60–3). The somite now has a more or less distinct triangular contour and consists of the dorsolateral dermomyotome and the ventromedial sclerotome. The tall columnar epithelial cell arrangement of the dermomyotome is evident. The entodermal cells of the midgut are simple squamous or simple cuboidal

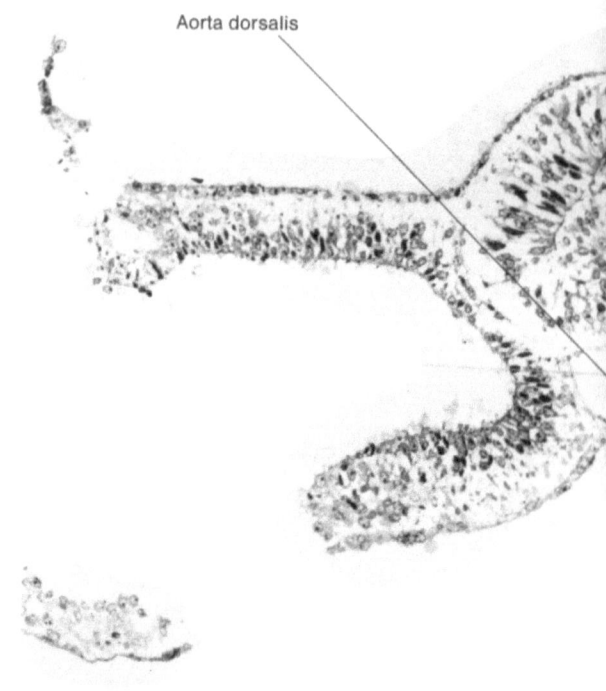

Aorta dorsalis

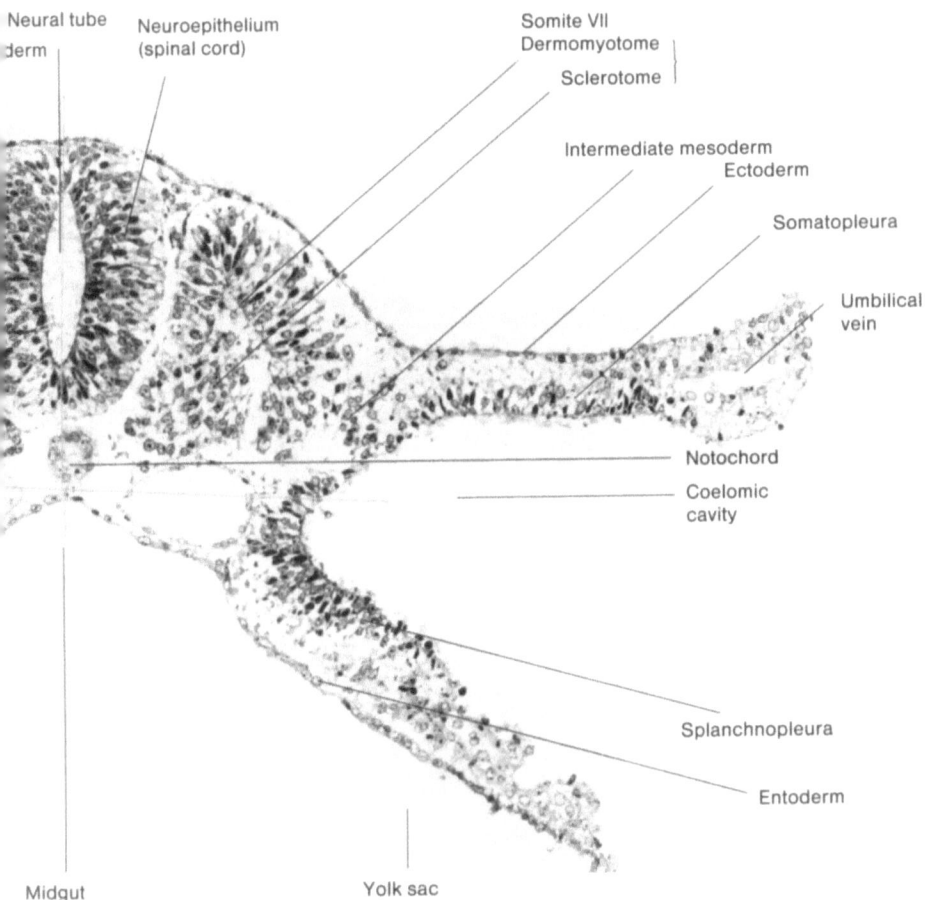

Neural tube
derm

Neuroepithelium
(spinal cord)

Somite VII
Dermomyotome

Sclerotome

Intermediate mesoderm

Ectoderm

Somatopleura

Umbilical
vein

Notochord

Coelomic
cavity

Splanchnopleura

Entoderm

Midgut

Yolk sac

Fig. 36. Section through seventh somite (66–2). The coelomic cavity widely communicates laterally with the extra-embryonic coelom. On the left side, between the ventrolateral aspect of the somite and the coelomic epithelium, a small cell group indicating the intermediate mesoderm appears for the first time

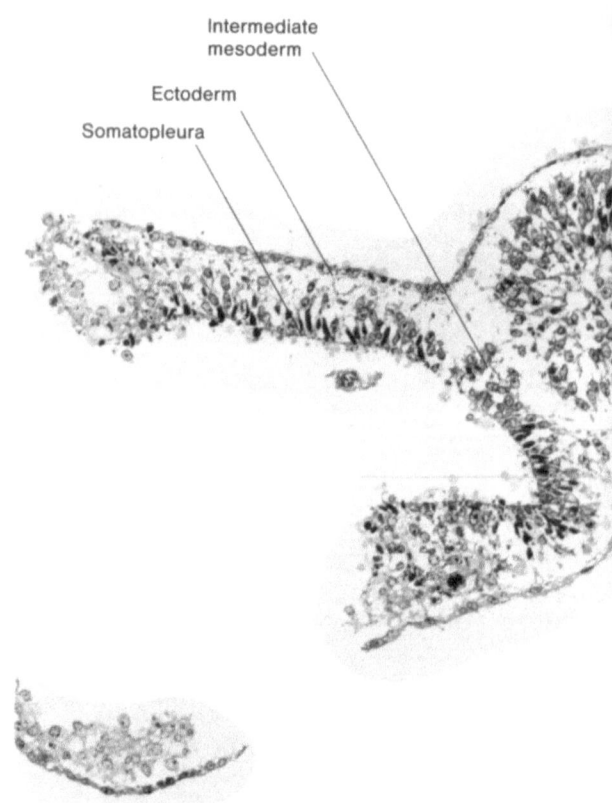

Intermediate
mesoderm

Ectoderm

Somatopleura

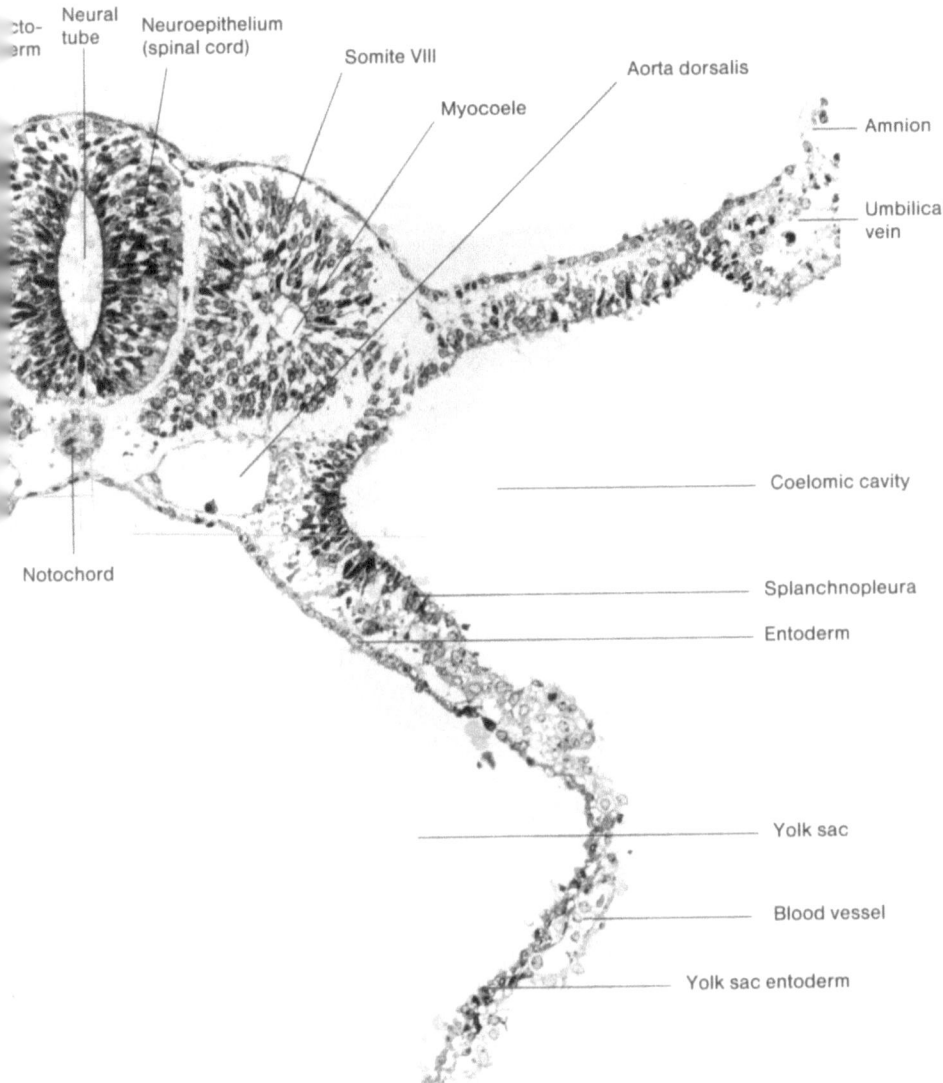

Fig. 37. Section through eighth somite (74–1). The somite is now quite large, shows a distinct triangular contour, and contains a lumen, the myocoele

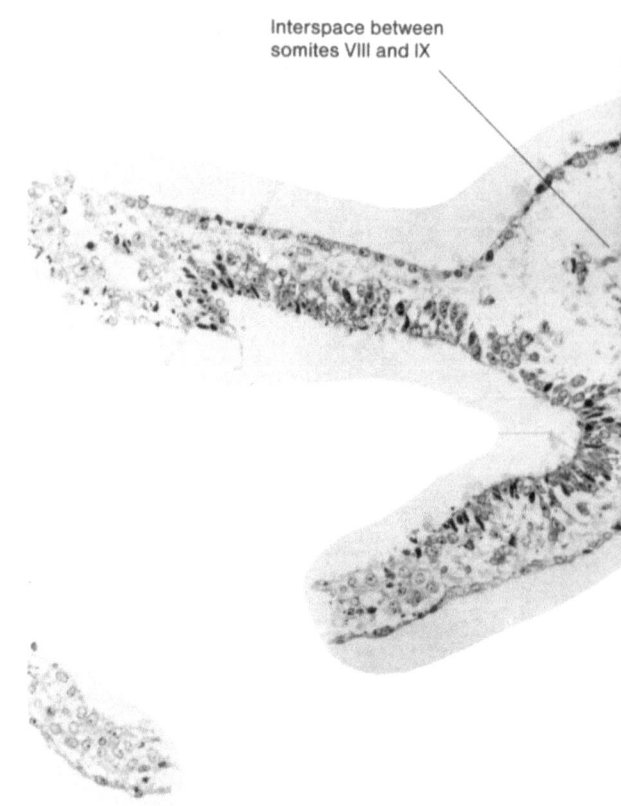

Interspace between
somites VIII and IX

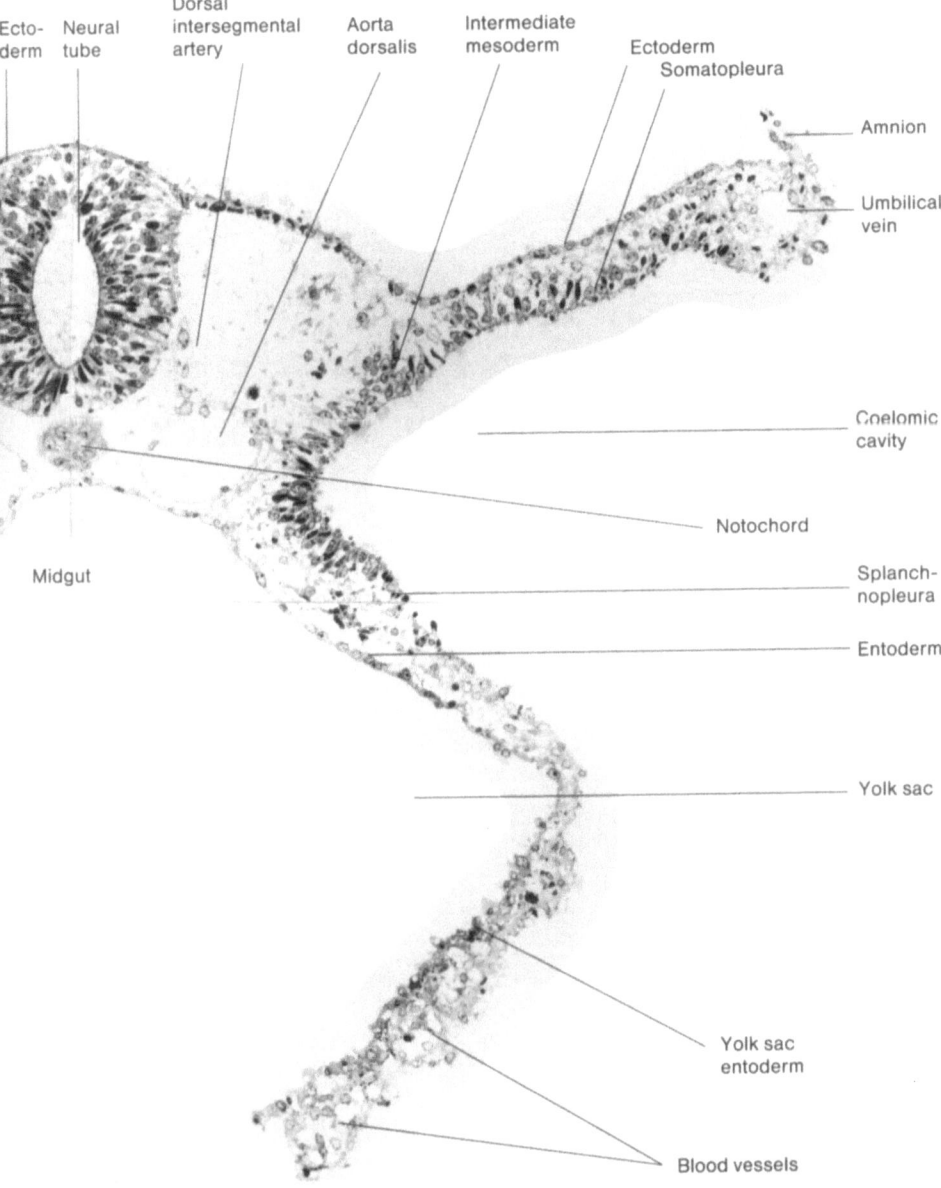

Ecto-derm | Neural tube | Dorsal intersegmental artery | Aorta dorsalis | Intermediate mesoderm | Ectoderm Somatopleura

Amnion

Umbilical vein

Coelomic cavity

Notochord

Midgut

Splanch-nopleura

Entoderm

Yolk sac

Yolk sac entoderm

Blood vessels

Fig. 38. Section through interspace between eighth and ninth somites (78–1). The interspace is almost cell-free except for a very small number of mesenchymal cells. On the left side sprouting of a dorsal intersegmental artery is clearly seen. The small cell group of the intermediate mesoderm is here half-embedded in the coelomic epithelium at the dorsomedial corner of the coelomic cavity on each side

Intermediate
mesoderm

Myocoele

Somite IX

Ectoderm

Ectoderm

Somatopleura

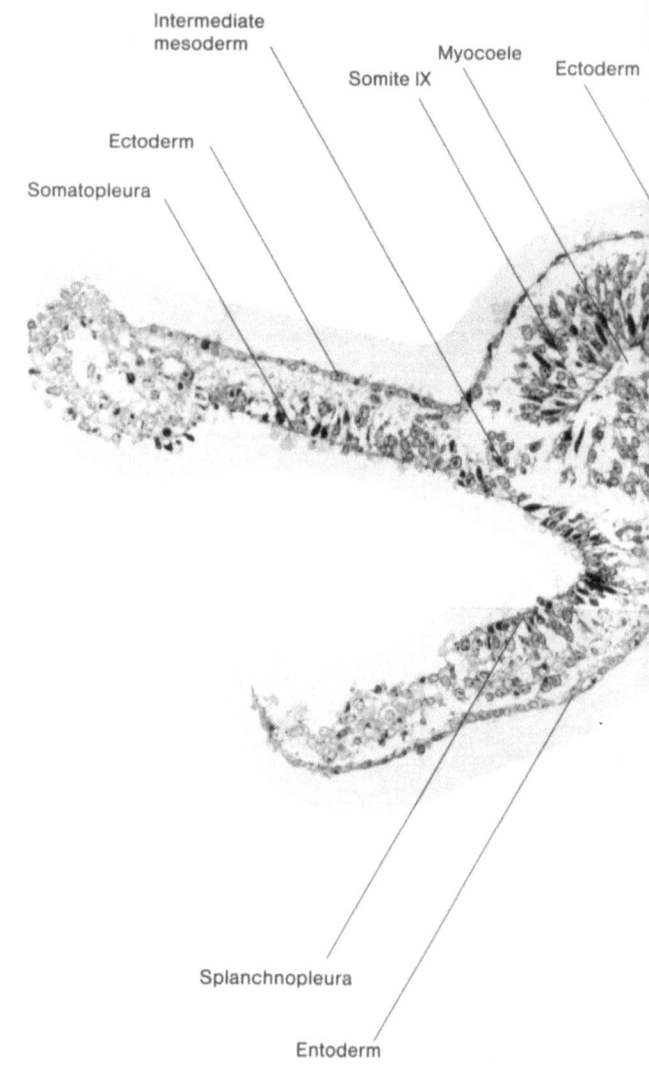

Splanchnopleura

Entoderm

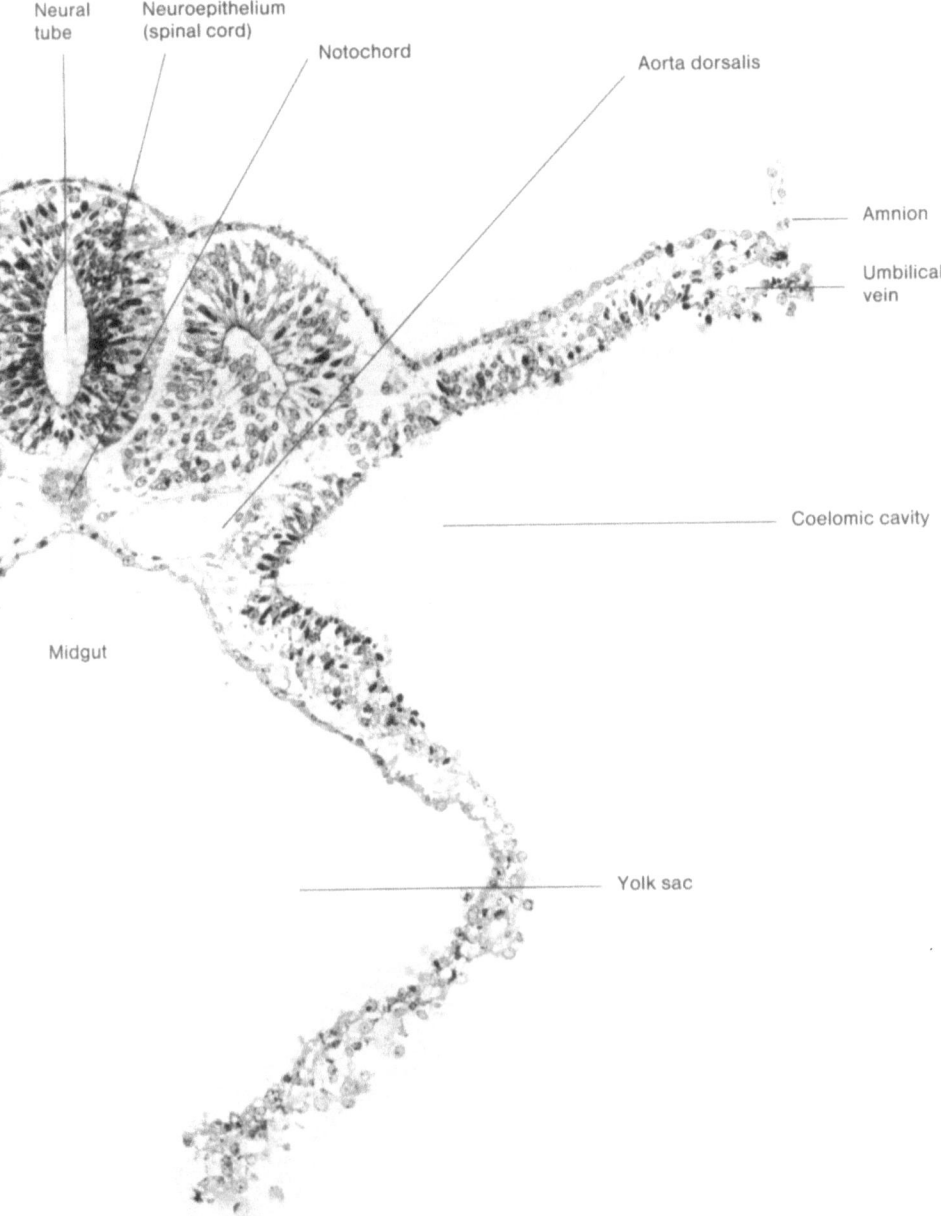

Neural tube

Neuroepithelium (spinal cord)

Notochord

Aorta dorsalis

Amnion

Umbilical vein

Coelomic cavity

Midgut

Yolk sac

Fig. 39. Section through ninth somite (82–1). The somite is large and contains a distinct myocoele. Also the sclerotomal cells are to some extent epithelial in character, especially in the dorsal half of the medial wall. The coelomic epithelium undergoes a sudden change into the loose mesenchymal tissue surrounding the yolk sac ventrally and underlining the amnion dorsolaterally

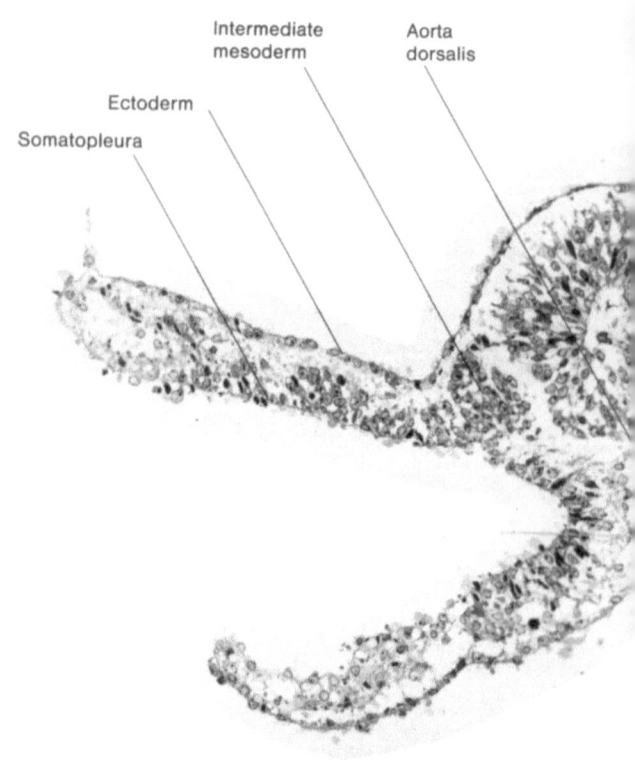

Somatopleura

Ectoderm

Intermediate
mesoderm

Aorta
dorsalis

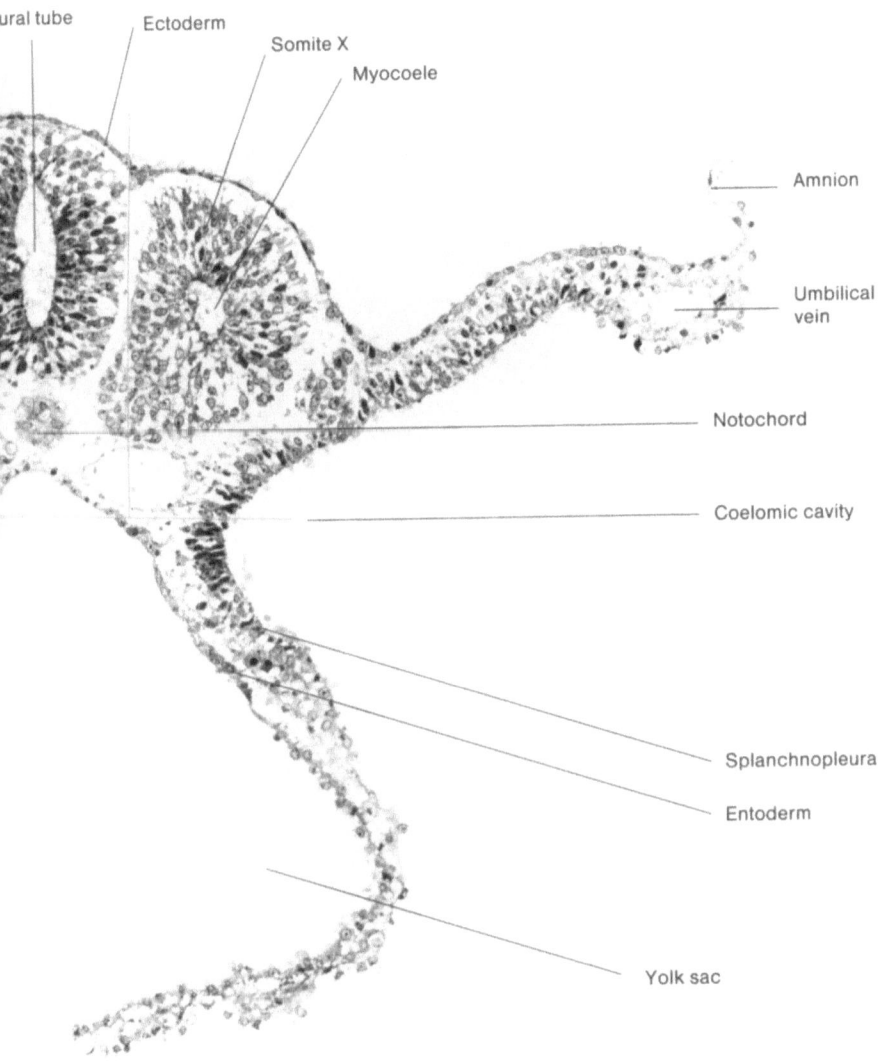

Ectoderm

Somite X

Myocoele

Amnion

Umbilical vein

Notochord

Coelomic cavity

Splanchnopleura

Entoderm

Yolk sac

Fig. 40. Section through tenth somite (90–3). The somite now shows a somewhat quadrangular contour and contains a distinct myocoele. Cells constituting the somite line up radially around the myocoele

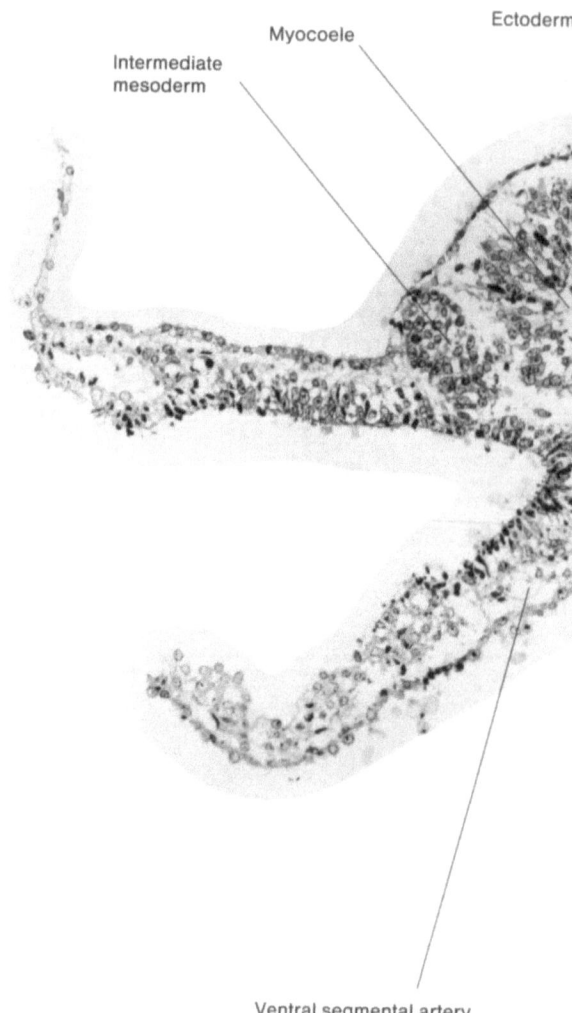

Intermediate mesoderm

Myocoele

Ectoderm

Ventral segmental artery

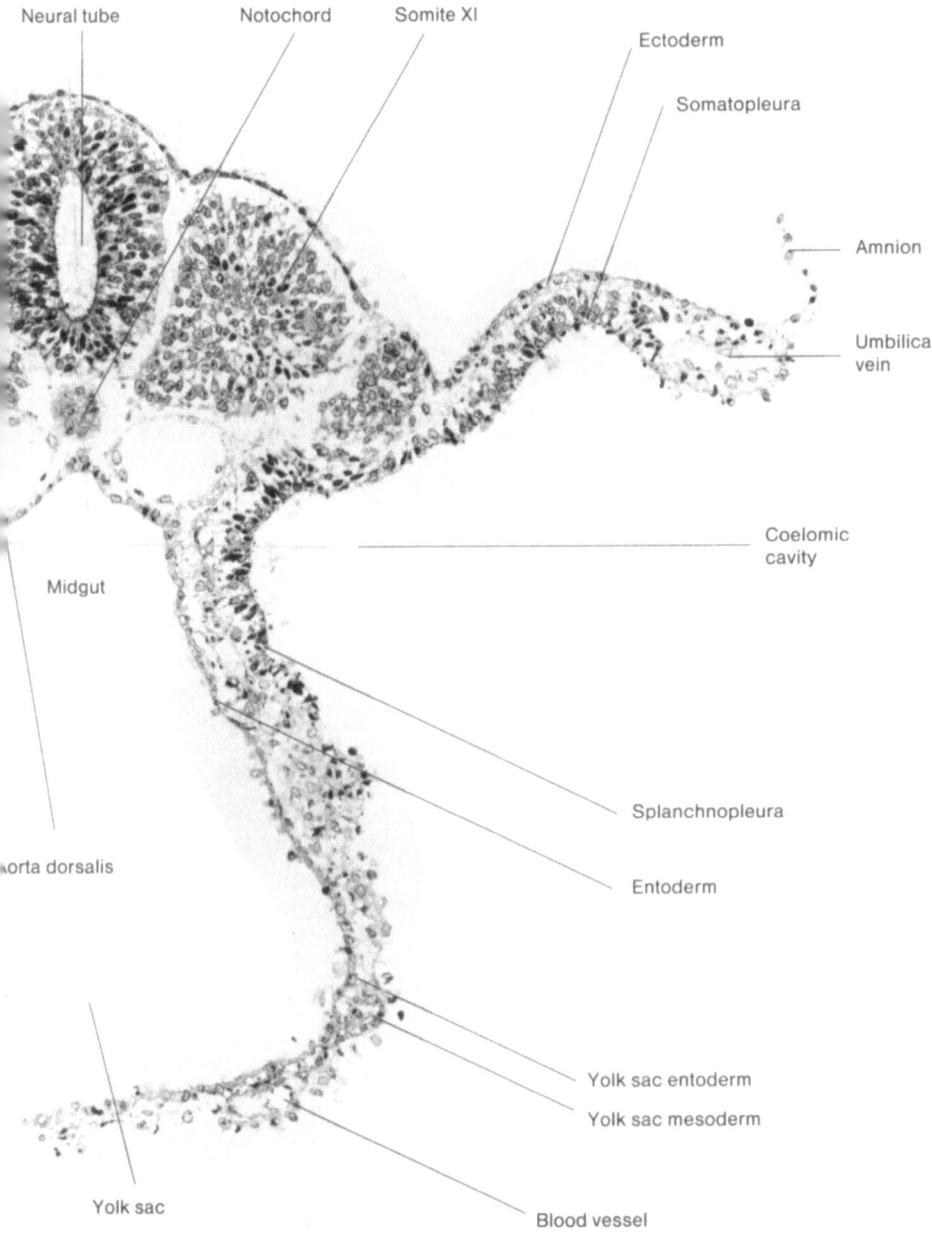

Neural tube   Notochord   Somite XI

Ectoderm

Somatopleura

Amnion

Umbilical
vein

Coelomic
cavity

Midgut

Splanchnopleura

Entoderm

aorta dorsalis

Yolk sac entoderm

Yolk sac mesoderm

Yolk sac

Blood vessel

Fig. 41. Section through 11th somite (98–1). The intermediate mesoderm increases markedly in size and shifts dorsolaterally, causing a faint bulge in the surface ectoderm above it. The sprouting of an omphalomesenteric artery from the dorsal aorta is evident on each side

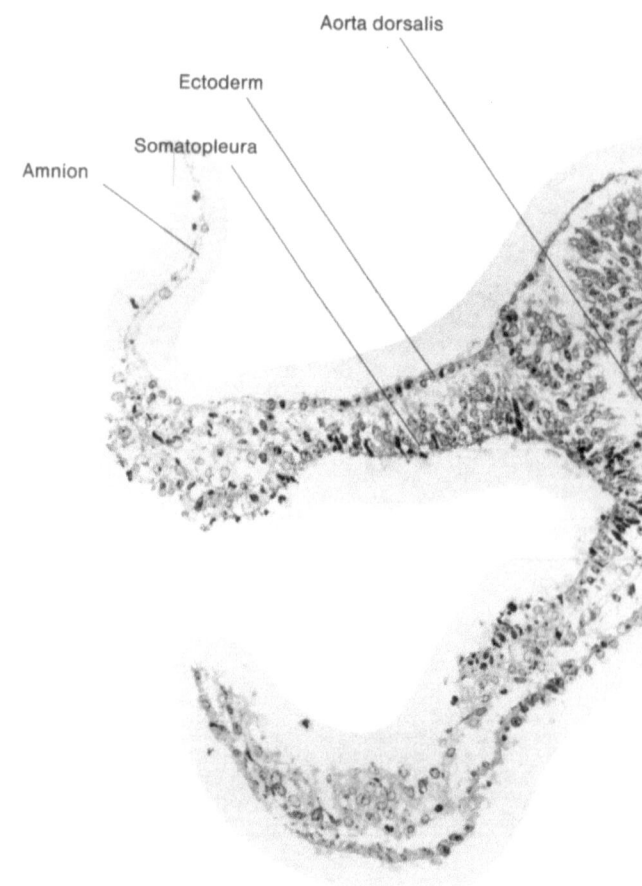

Amnion

Somatopleura

Ectoderm

Aorta dorsalis

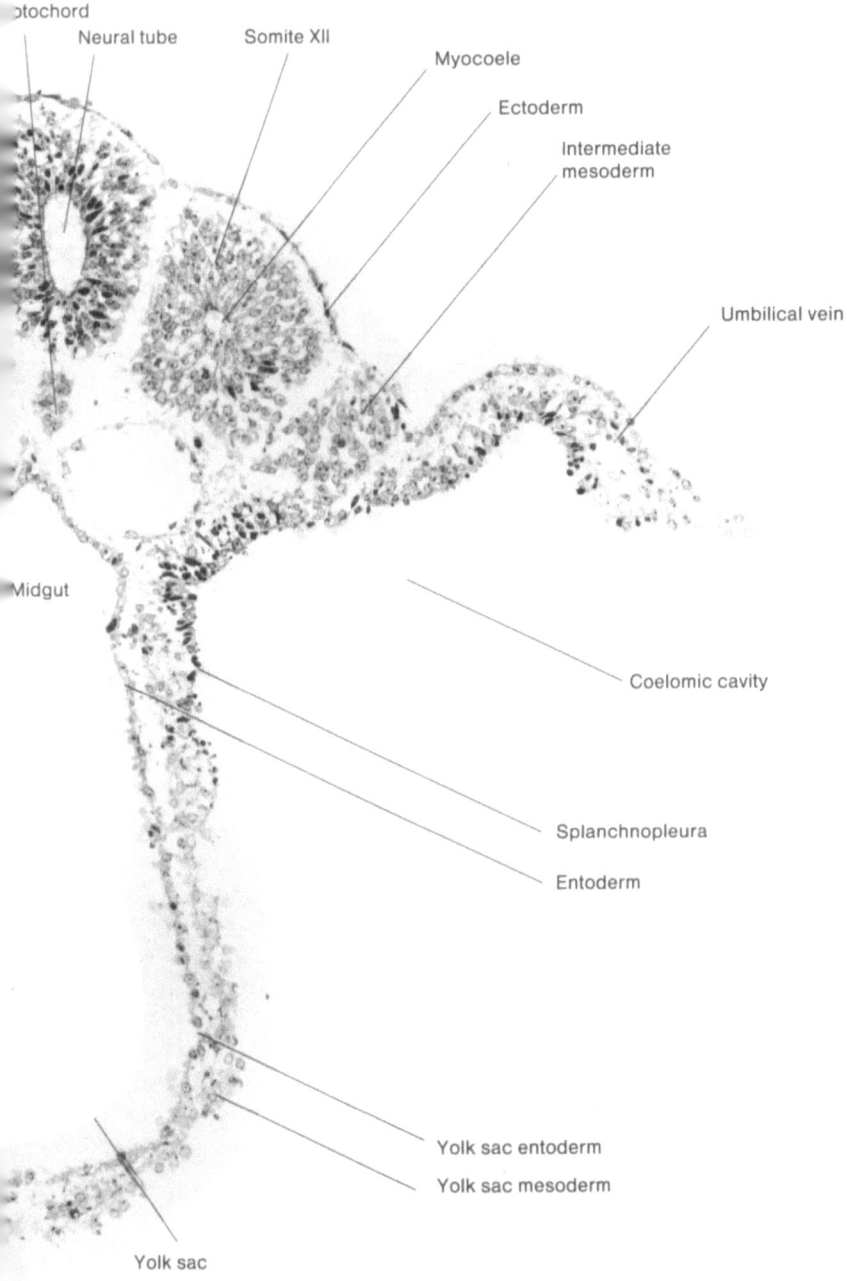

Notochord

Neural tube   Somite XII

Myocoele

Ectoderm

Intermediate
mesoderm

Umbilical vein

Midgut

Coelomic cavity

Splanchnopleura

Entoderm

Yolk sac entoderm

Yolk sac mesoderm

Yolk sac

Fig. 42. Section through 12th somite (105–3). The yolk sac together with the midgut becomes narrower, indicating the approach of the posterior intestinal portal. The dorsal aortae on each side approach each other but do not touch at any point. The intermediate mesoderm attains its maximum size but shows no sign of pronephric differentiation

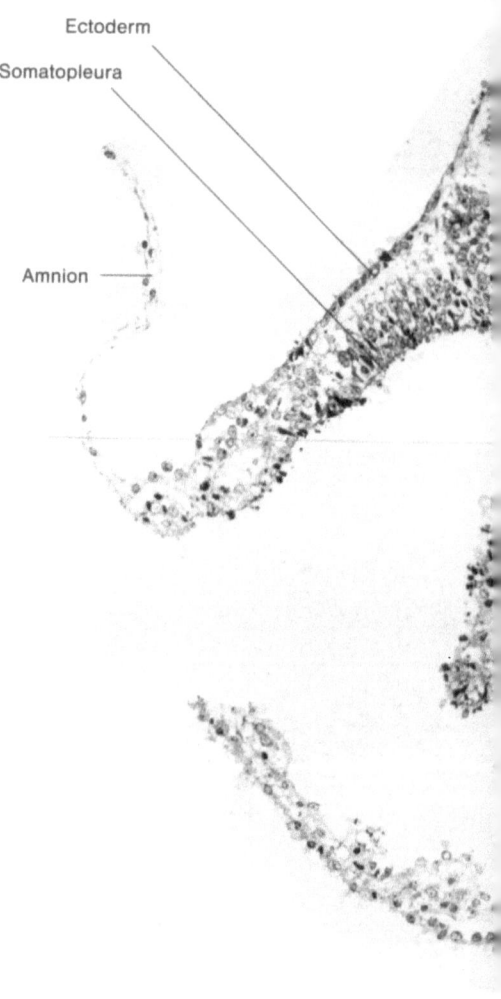

Ectoderm

Somatopleura

Amnion

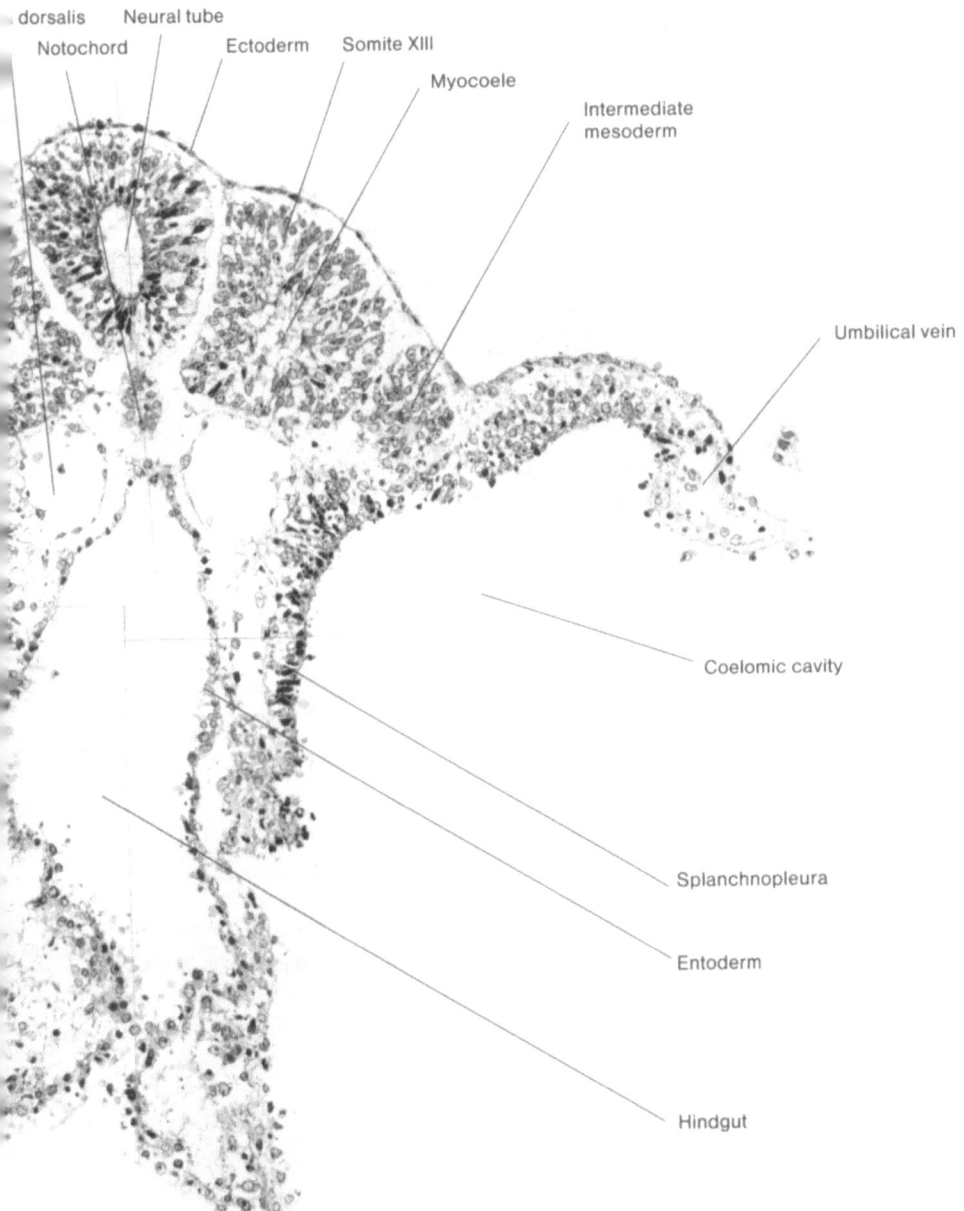

dorsalis   Neural tube
Notochord   Ectoderm   Somite XIII
                              Myocoele
                                        Intermediate
                                        mesoderm

Umbilical vein

Coelomic cavity

Splanchnopleura

Entoderm

Hindgut

Fig. 43. Section through 13th somite and posterior intestinal portal (112–2). The right and left walls of the yolk sac touch each other and separate the intestine from the yolk sac, and the midgut passes here on to the hindgut. The embryonic body shows now again a more or less vertical configuration. The quadrangular somite with a distinct myocoele and the cell aggregation of the intermediate mesoderm are conspicuous

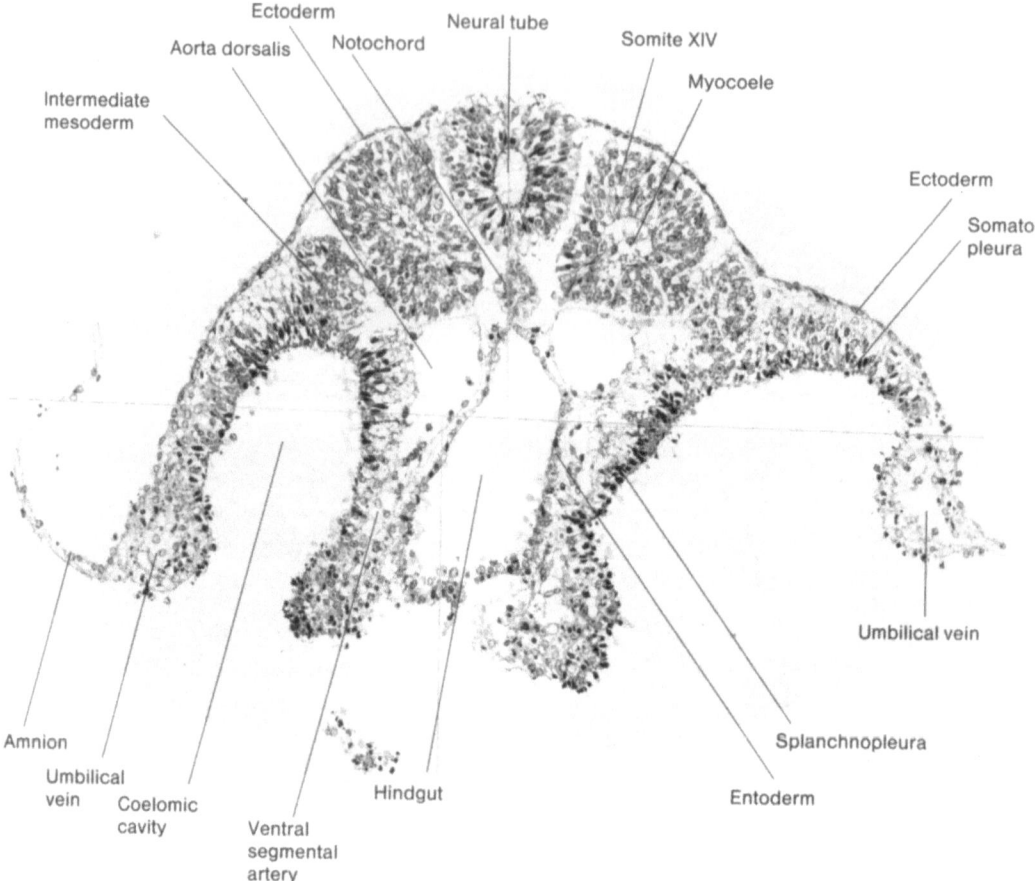

Ectoderm
Aorta dorsalis
Notochord
Neural tube
Somite XIV
Myocoele
Intermediate
mesoderm
Ectoderm
Somato-
pleura
Umbilical vein
Amnion
Splanchnopleura
Umbilical
vein
Coelomic
cavity
Hindgut
Entoderm
Ventral
segmental
artery

Fig. 44. Section through 14th somite and anterior end of hindgut (118–1). The dorsal half of the hindgut is lined by a simple squamous, whereas the ventral half by a simple cuboidal entodermal epithelium. The surface ectoderm covering the coelomic cavity shifts ventrally so that it envelops more and more the ventral half of the body. The somite and intermediate mesoderm are conspicuous, as in Fig. 43. Sprouting of an omphalomesenteric artery from the dorsal aorta is evident on the right side

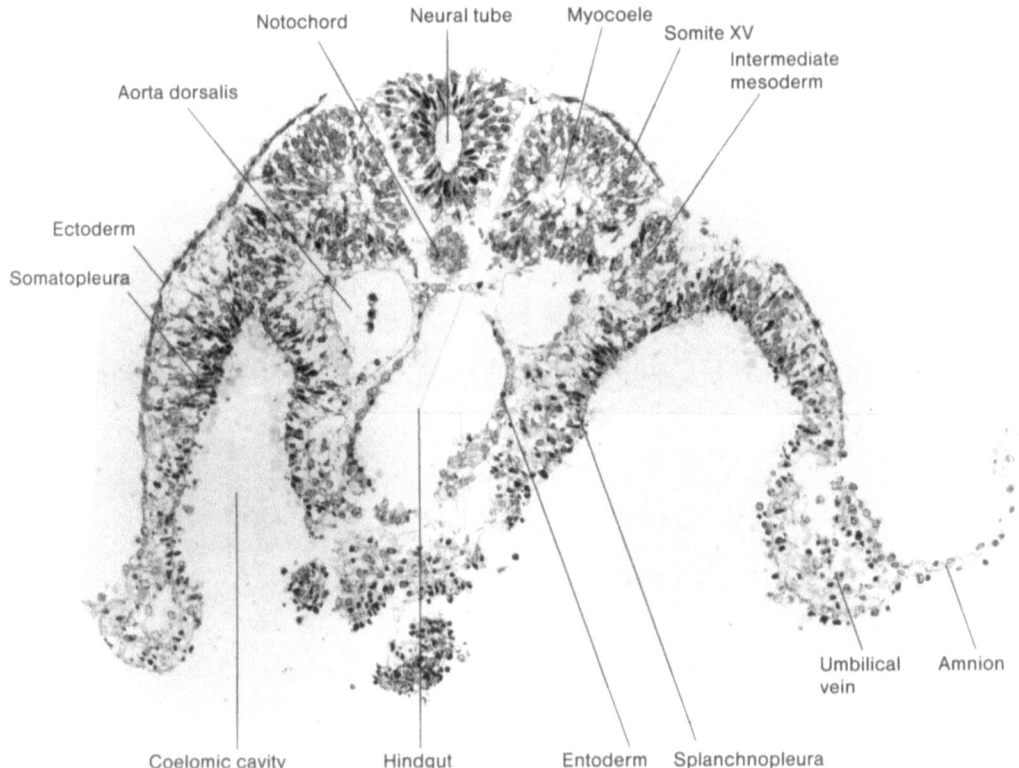

Fig. 45. Section through 15th somite (124). The enveloping of the ventral half of the body by the surface ectoderm has progressed. The neural tube becomes smaller but maintains a distinct oval configuration and consists of a pseudostratified columnar epithelium with three to four rows of nuclei. The somite and intermediate mesoderm are prominent

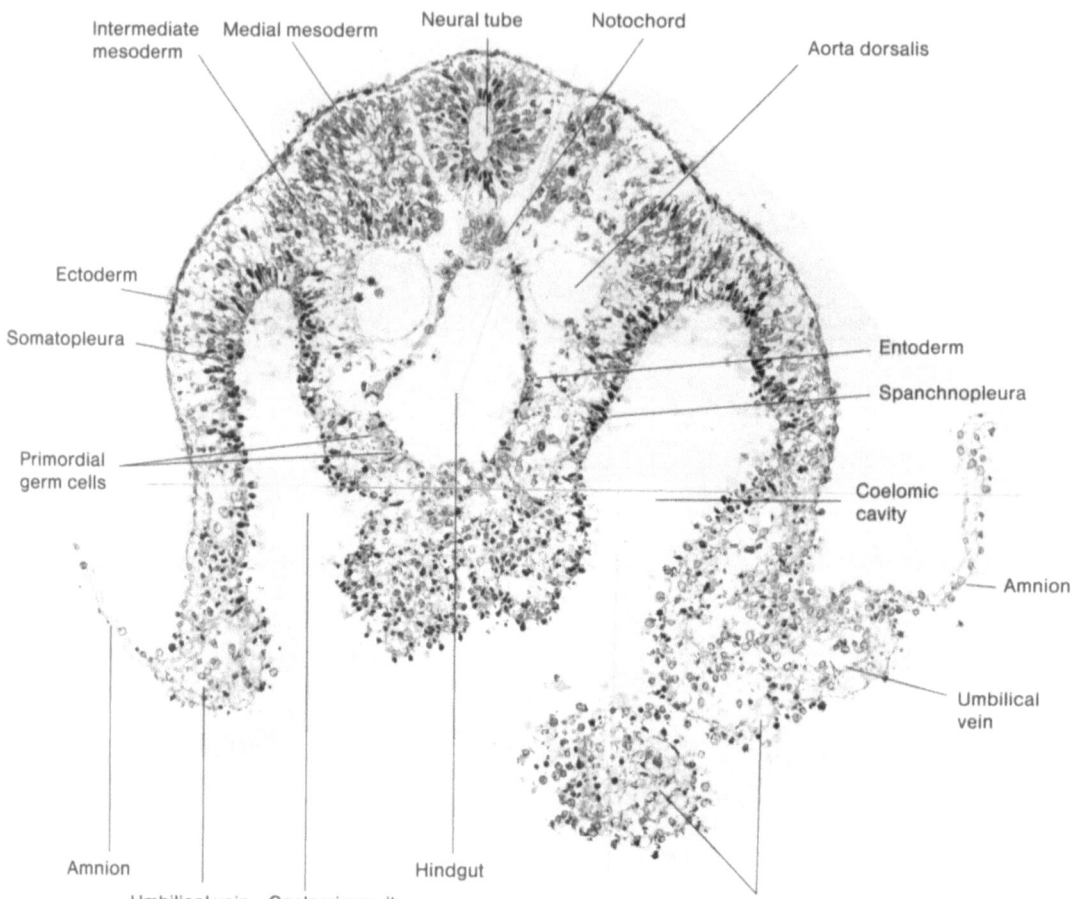

Intermediate mesoderm

Medial mesoderm

Neural tube

Notochord

Aorta dorsalis

Ectoderm

Somatopleura

Primordial germ cells

Entoderm

Spanchnopleura

Coelomic cavity

Amnion

Umbilical vein

Amnion

Umbilical vein   Coelomic cavity

Hindgut

Body stalk

Fig. 46. Section through interspace between 15th somite and non-segmented medial meso-derm (129–3). The interspace is seen on the left side, whereas on the right side the non-segmented medial mesoderm appears. Among the entodermal cells lining the ventral half of the hindgut numerous large round lucent cells are found; they represent the primordial germ cells

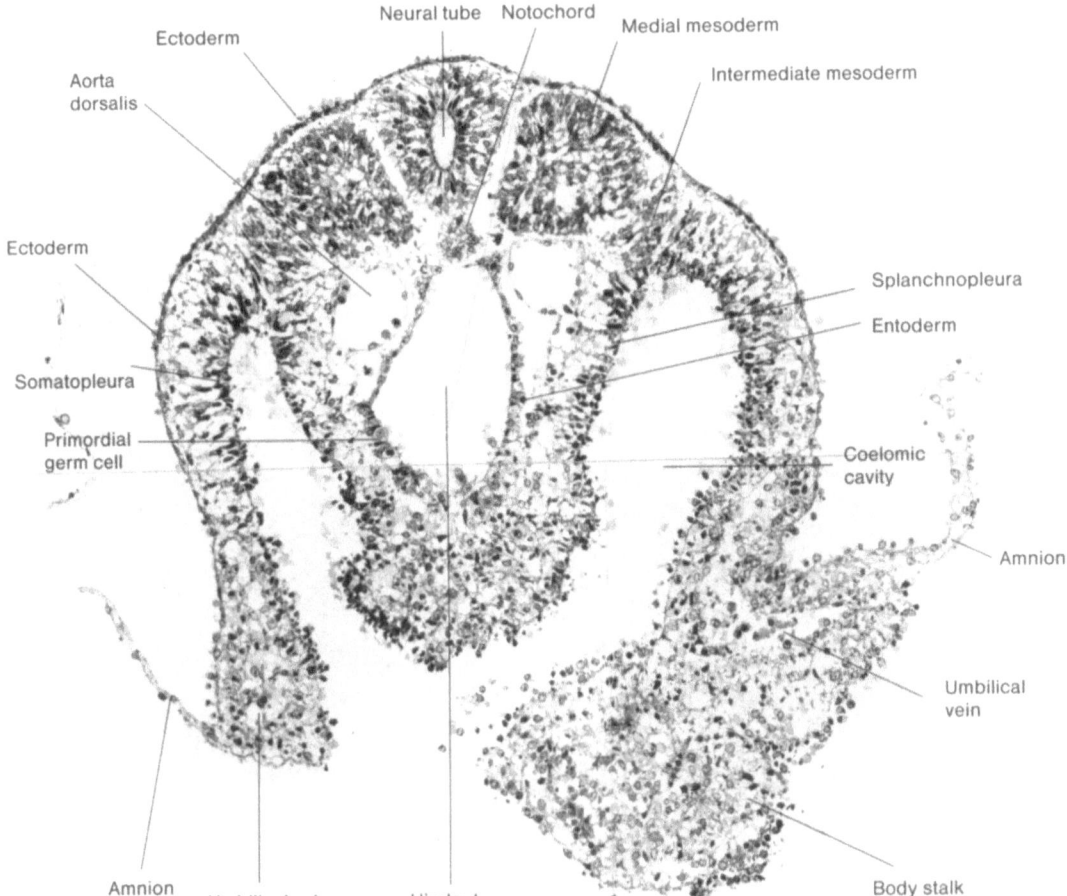

Ectoderm Neural tube Notochord Medial mesoderm Intermediate mesoderm Aorta dorsalis Ectoderm Somatopleura Primordial germ cell Splanchnopleura Entoderm Coelomic cavity Amnion Umbilical vein Amnion Umbilical vein Hindgut Body stalk

Fig. 47. Section through anterior portion of body stalk (131–3). The enveloping of the body by the surface ectoderm progresses so that the right and left coelomic cavities join each other ventral to the hindgut and then open ventrally into the extra-embryonic coelom. On the left side a massive mesenchymal tissue of the body stalk adheres to the ventral aspect of the amnio-ectodermal junction where the umbilical vein turns laterally and comes into the body stalk. The primordial germ cells are conspicuous also in this section among the entodermal epithelial cells of the ventral half of the hindgut

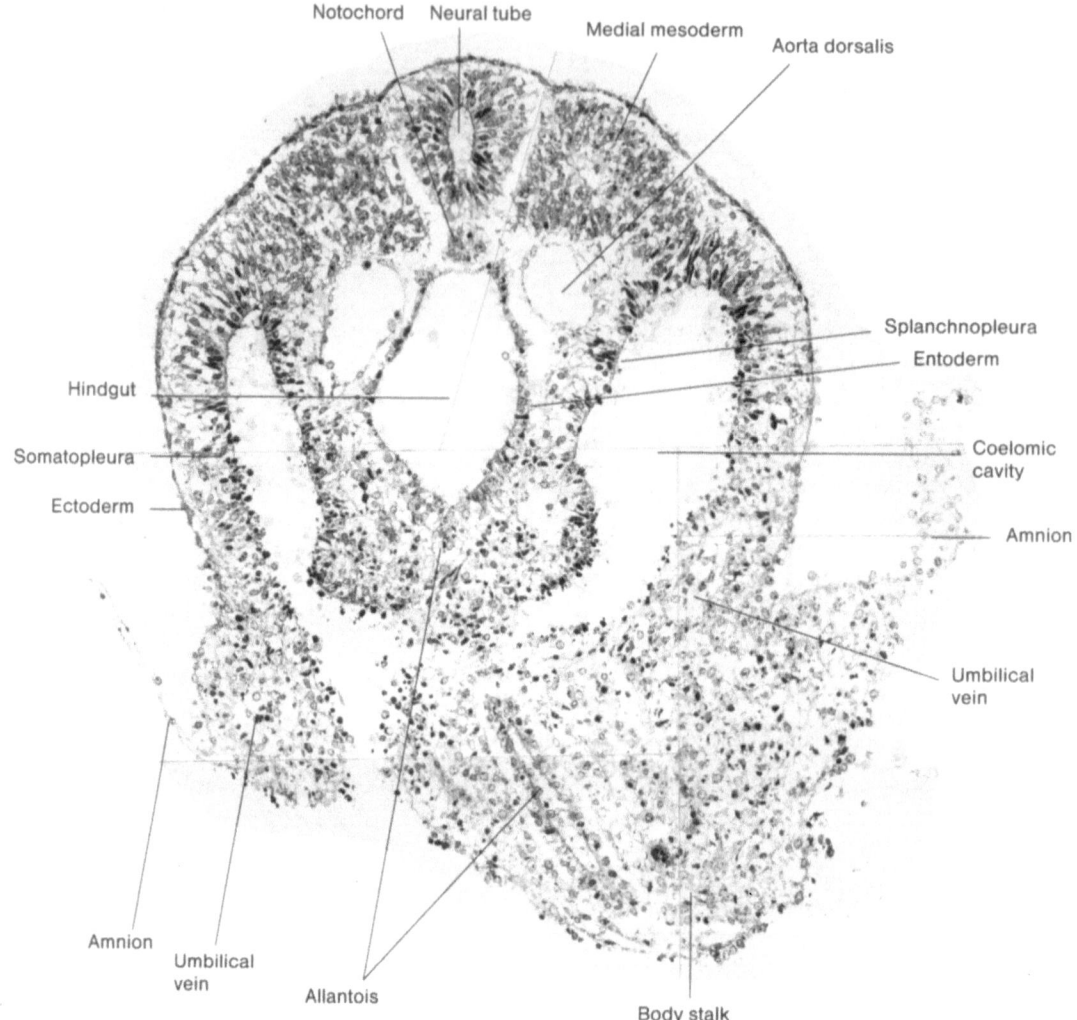

Notochord   Neural tube   Medial mesoderm   Aorta dorsalis

Splanchnopleura
Entoderm

Hindgut
Somatopleura
Ectoderm

Coelomic
cavity

Amnion

Umbilical
vein

Amnion   Umbilical   Allantois   Body stalk
vein

Fig. 48. Section through beginning of allantois (134–1). From the ventral extremity of the hindgut sprouts a small entodermal bud, and ventral to it a tubular structure is embedded in the mesenchymal tissue of the body stalk. They represent the origin and the initial portion of the allantois respectively. The non-segmented medial mesoderm adjoins the coelomic epithelium. The coelomic cavity on the left side is just closed by the mesenchymal tissue of the body stalk, whereas on the right side it still keeps a narrow communication with the extra-embryonic coelom

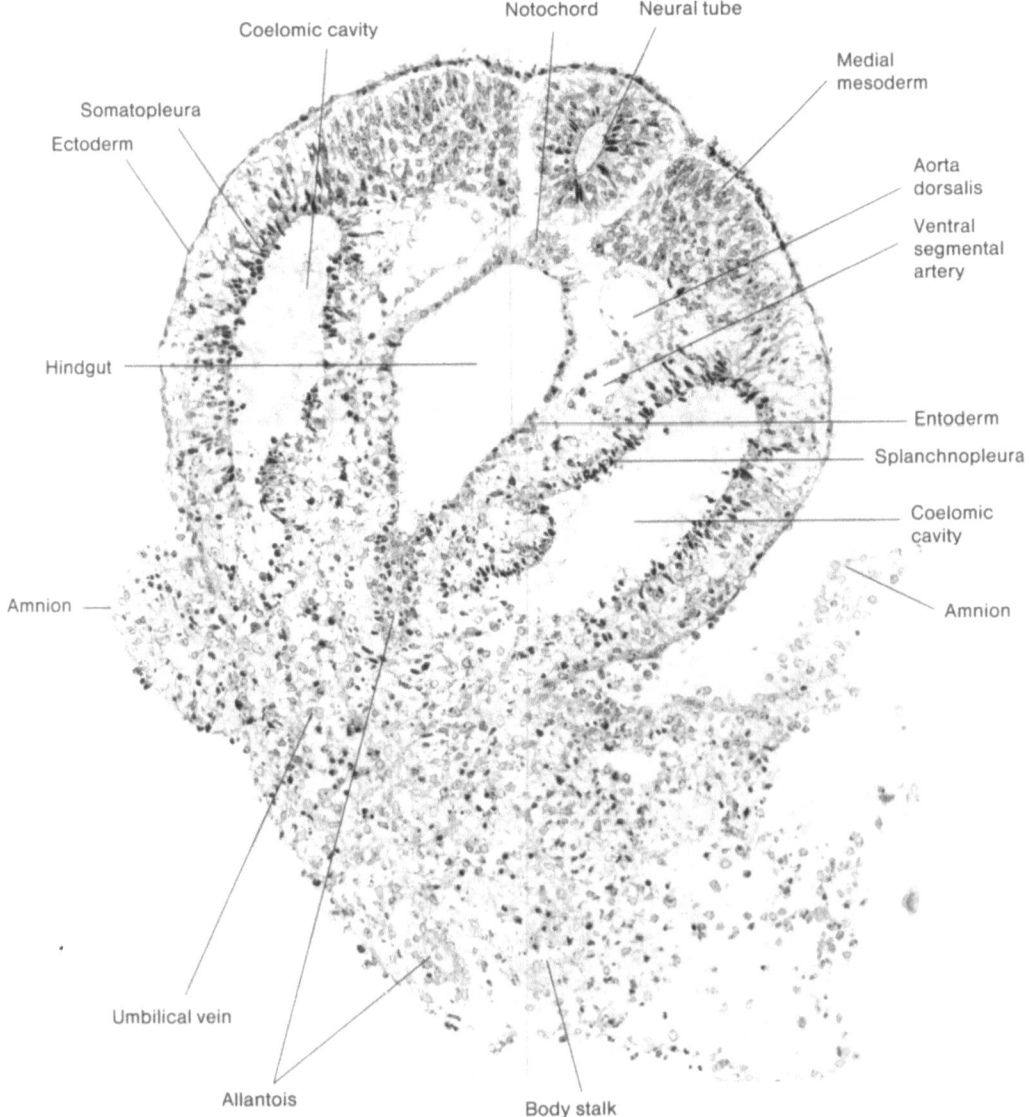

Fig. 49. Section through level where body stalk connects with body (137–1). The ventral aspect of the body now connects with the massive mesenchymal tissue of the body stalk, in which the right umbilical vein and a section of the allantois are recognized. From the ventral extremity of the hindgut sprouts the allantois. The coelomic cavities lateral to the hindgut are closed ventrally by the body stalk tissue. The ventral segmental artery sprouting from the dorsal aorta is seen on each side

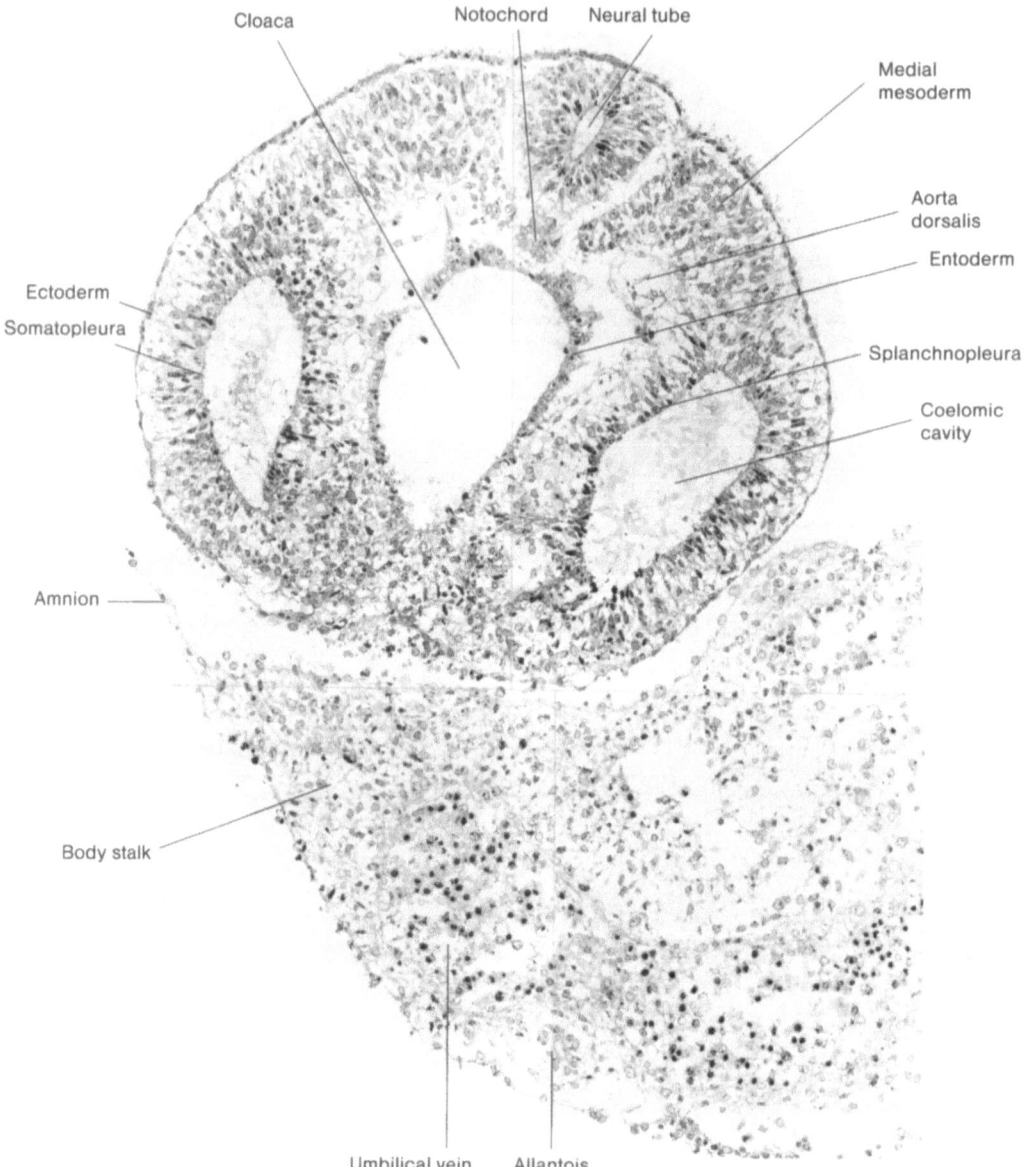

Fig. 50. Section through level where body stalk separates again from body (145–1). The surface ectoderm envelops the body completely. At the portion corresponding to the ventral extremity of the hindgut, now the cloaca, it makes a shallow depression indicating the proctodeum. The neural tube still consists of pseudostratified columnar epithelium with three to five rows of nuclei. In the body stalk a large umbilical vein is evident. The mesodermal epithelial cells lining the coelomic cavity are dispersing radially everywhere to underline the surface ectoderm and the entoderm of the cloaca

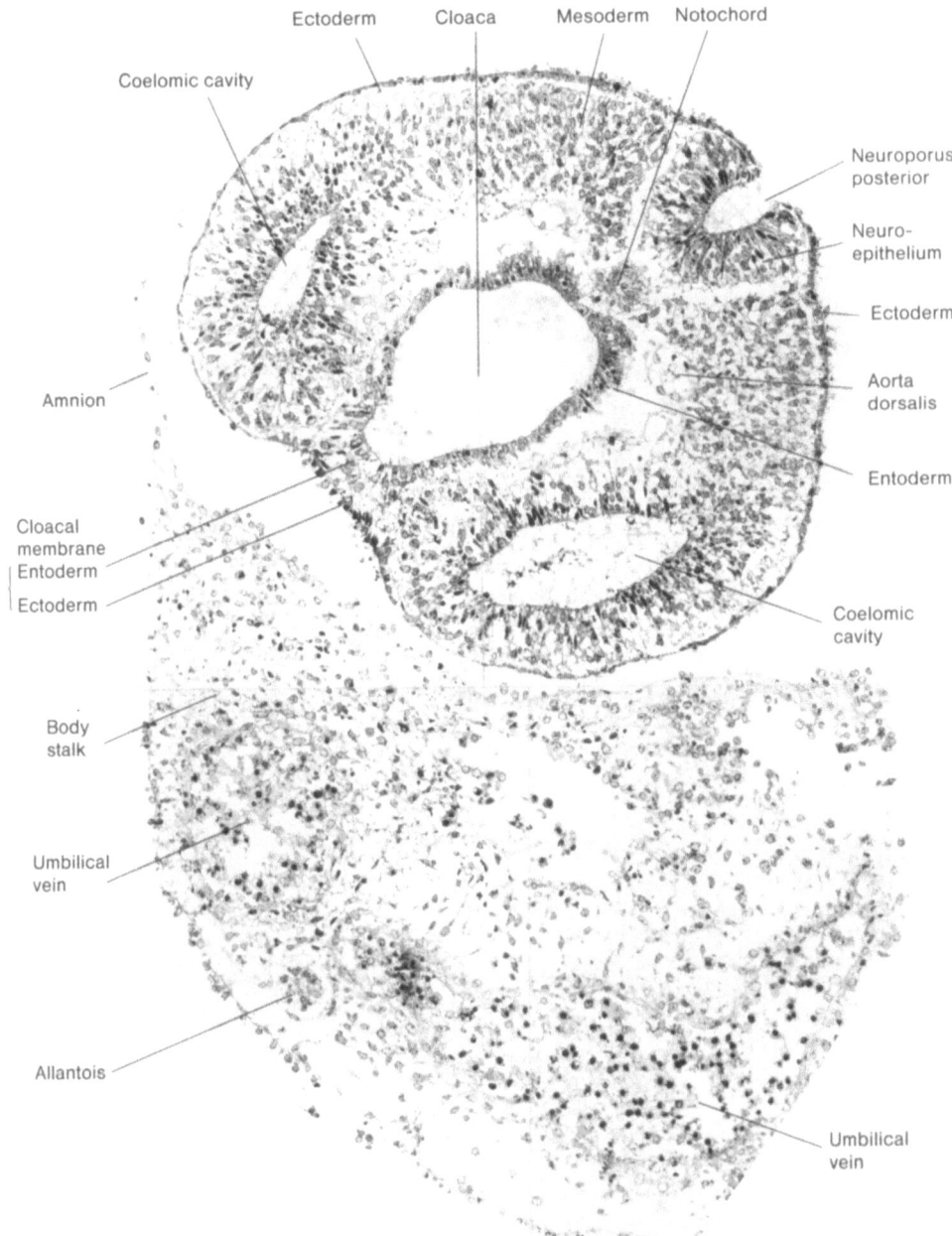

Fig. 51. Section through posterior neuropore and anterior end of cloacal membrane (148–2). The neural tube opens again dorsally into the amniotic cavity as the posterior neuropore. The cloaca is now relatively large, and at its ventral extremity the entodermal cells come in contact with the surface ectodermal cells, making a two-cell-layer membrane, the cloacal membrane. From now on the coelomic cavity rapidly diminishes in size

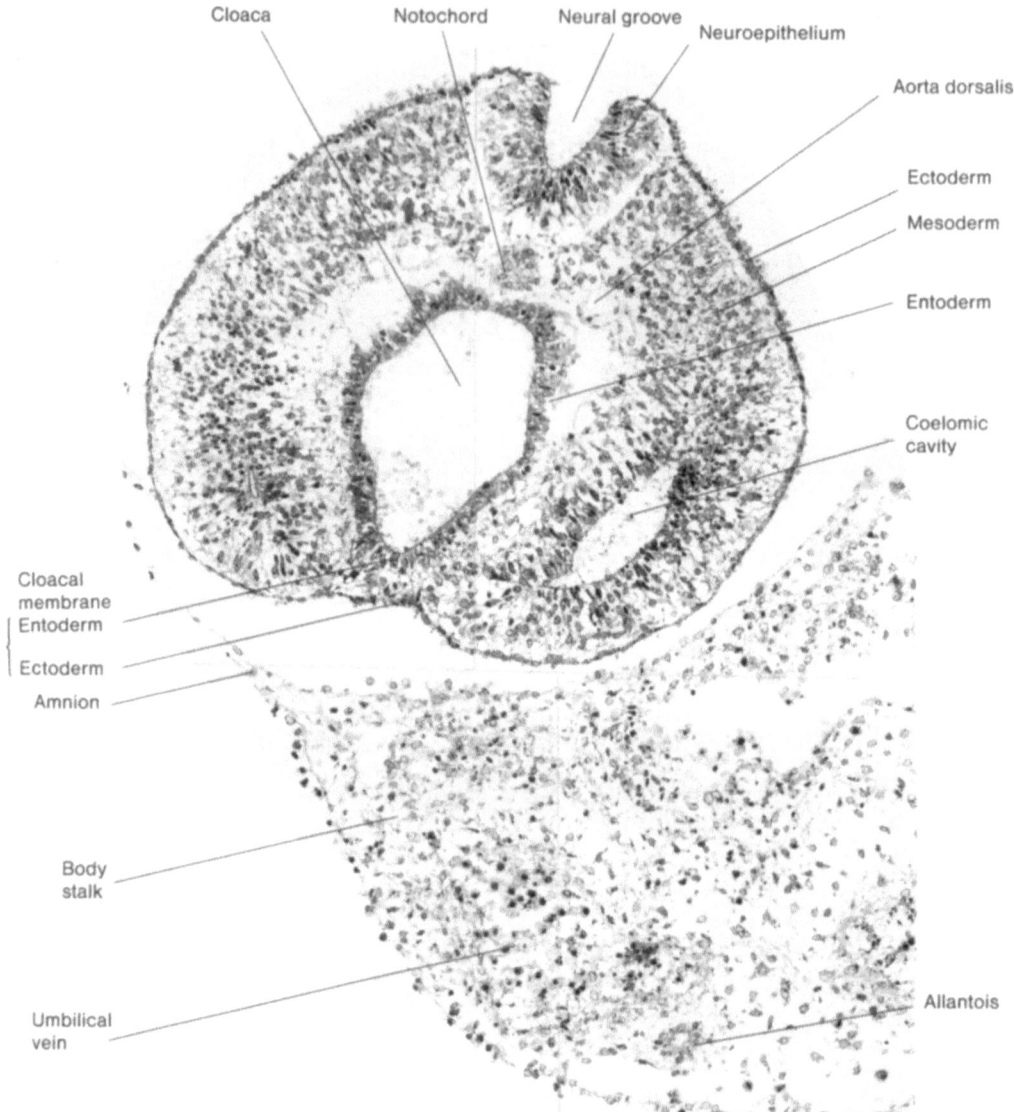

Cloaca

Notochord

Neural groove

Neuroepithelium

Aorta dorsalis

Ectoderm

Mesoderm

Entoderm

Coelomic
cavity

Cloacal
membrane
Entoderm

Ectoderm

Amnion

Body
stalk

Umbilical
vein

Allantois

Fig. 52. Section through middle portion of cloacal membrane (150–1). The right coelomic cavity disappears, whereas the left one remains as a small cavity. The opening of the neural groove widens, and the neural plate itself slightly increases in size. The allantois shifts left-ventrally in the body stalk

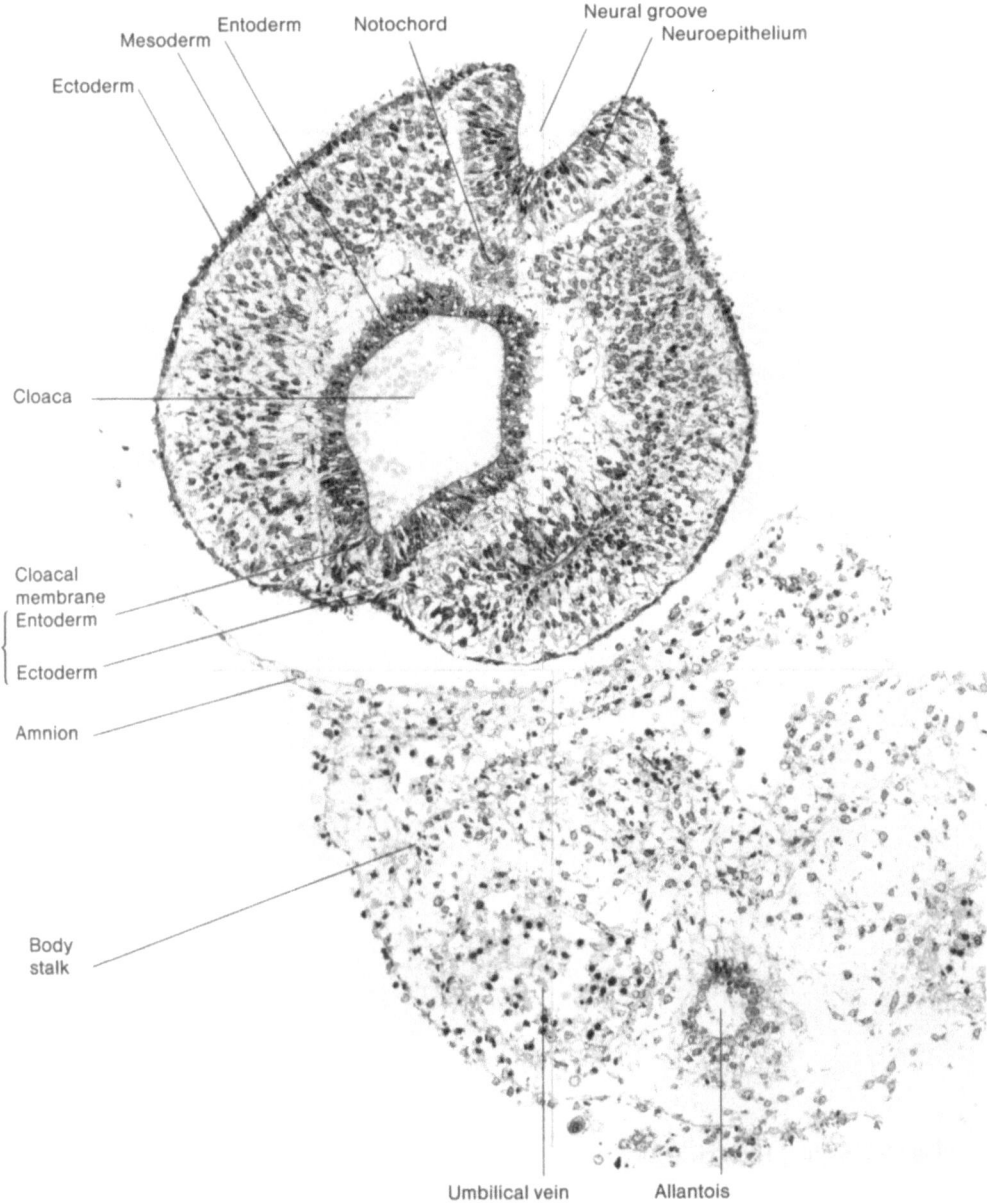

Ectoderm

Mesoderm Entoderm Notochord Neural groove
Neuroepithelium

Cloaca

Cloacal
membrane
Entoderm

Ectoderm

Amnion

Body
stalk

Umbilical vein Allantois

Fig. 53. Section through posterior end of coelomic cavity (153–1). The left coelomic cavity disappears. The large cloaca is lined by tall columnar epithelium and the cloacal membrane consists of the columnar entodermal cells and simple squamous ectodermal cells. The section of the allantois increases in size, indicating the allantoic sac

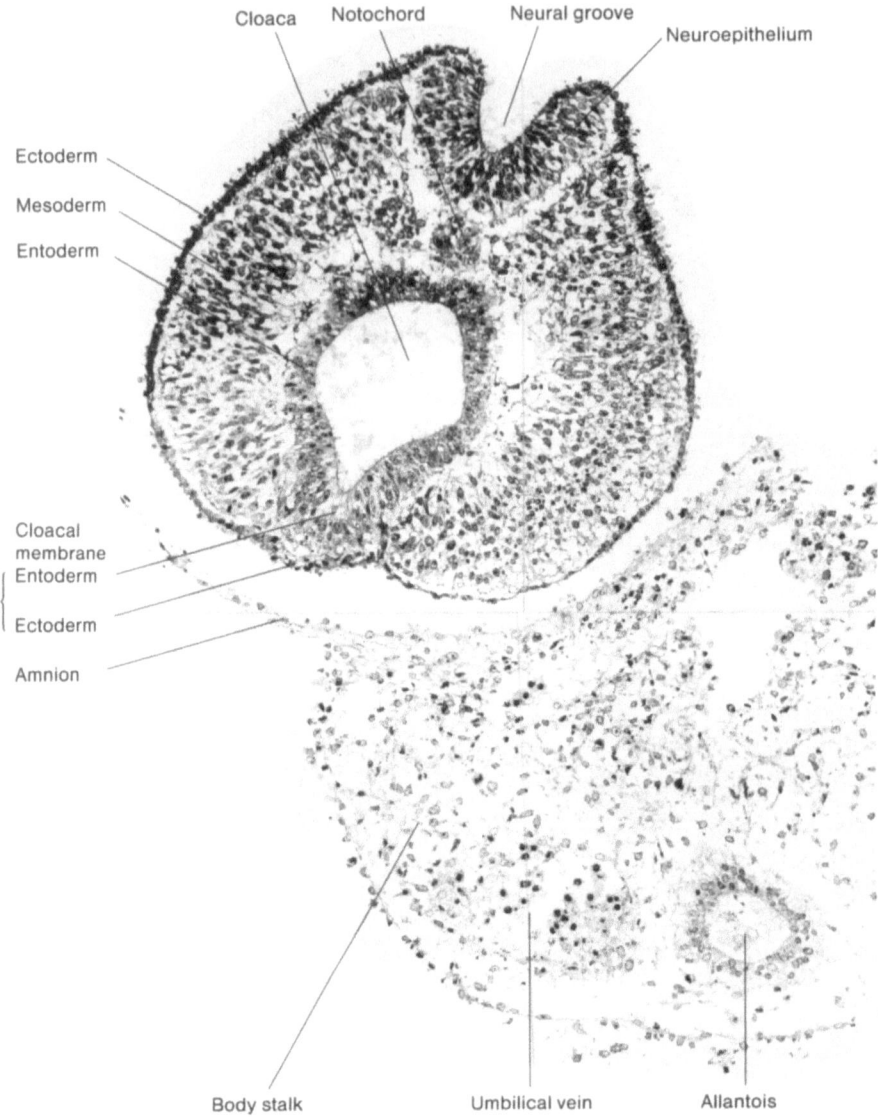

Labels in figure:
Cloaca · Notochord · Neural groove · Neuroepithelium
Ectoderm · Mesoderm · Entoderm
Cloacal membrane Entoderm · Ectoderm · Amnion
Body stalk · Umbilical vein · Allantois

Fig. 54. Section through posterior end of cloacal membrane (155–1). The two-cell-layer cloacal membrane consisting of tall columnar entodermal cells and simple squamous ectodermal cells is obvious. The wide space between the surface ectoderm and cloaca is filled with randomly dispersed mesenchymal cells. The neural groove becomes shallower. In the body stalk the allantoic sac is seen

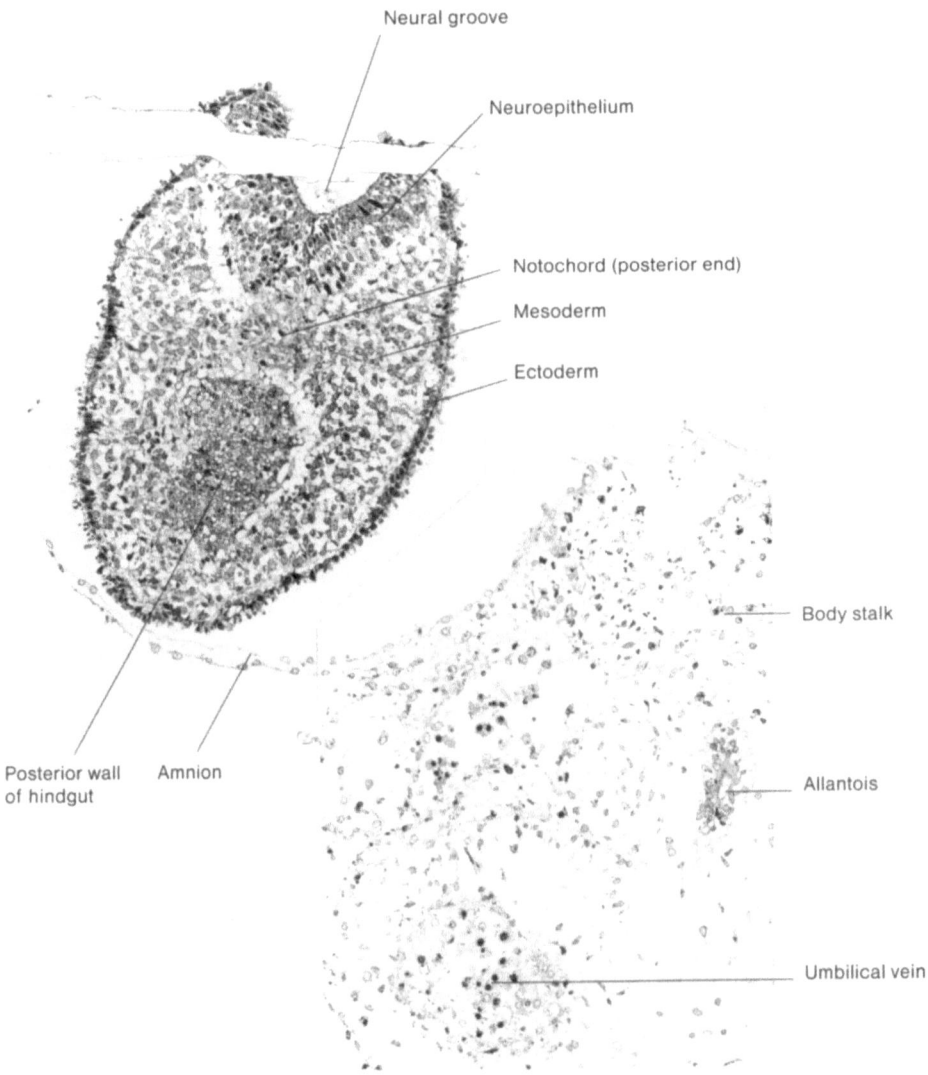

Neural groove

Neuroepithelium

Notochord (posterior end)

Mesoderm

Ectoderm

Body stalk

Posterior wall   Amnion
of hindgut

Allantois

Umbilical vein

Fig. 55. Section through posterior end of notochord and of hindgut (cloaca) (156–2). Cells of the notochord are intermingling dorsally with those of the neural plate, which increases in size and becomes shallower. An oval area of densely packed cells ventral to the notochord indicates the tangential section of the posterior wall of the cloaca

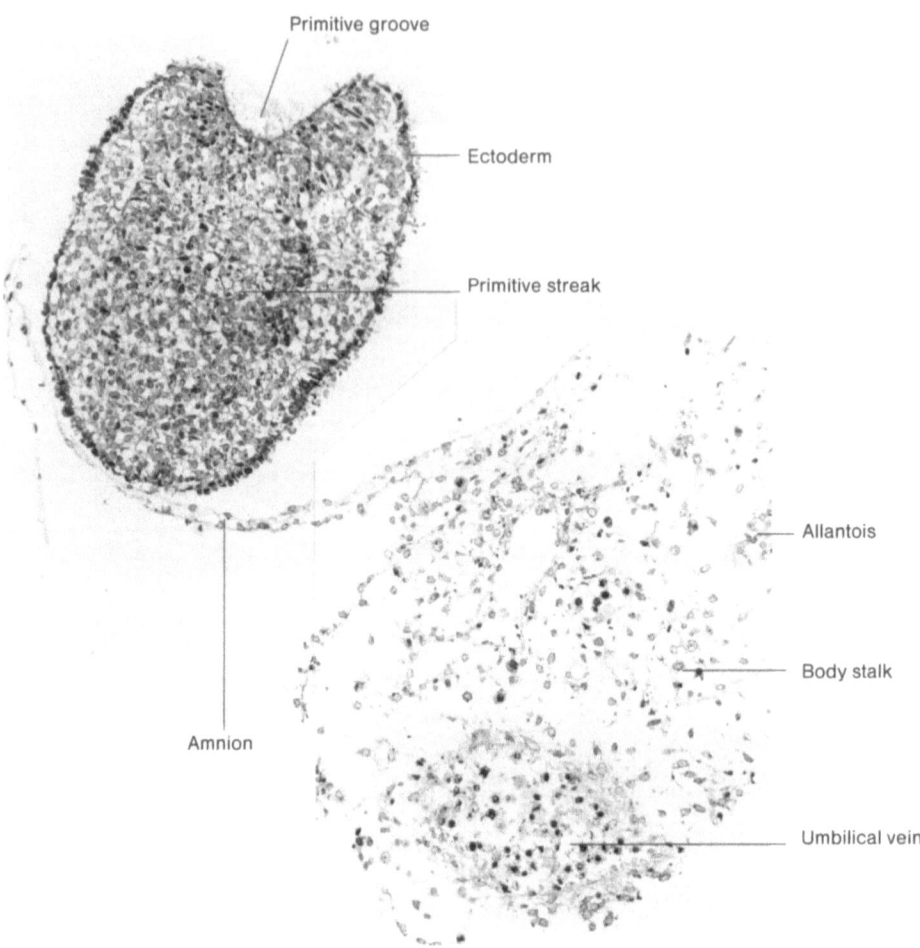

Fig. 56. Section through primitive groove and primitive streak (160–3). The notochord can no longer be identified. Ventral to the neural plate there is a distinct area consisting of densely packed cells; this represents the primitive streak. The neural plate loses the basal limiting membrane and its cells also participate in forming the primitive streak

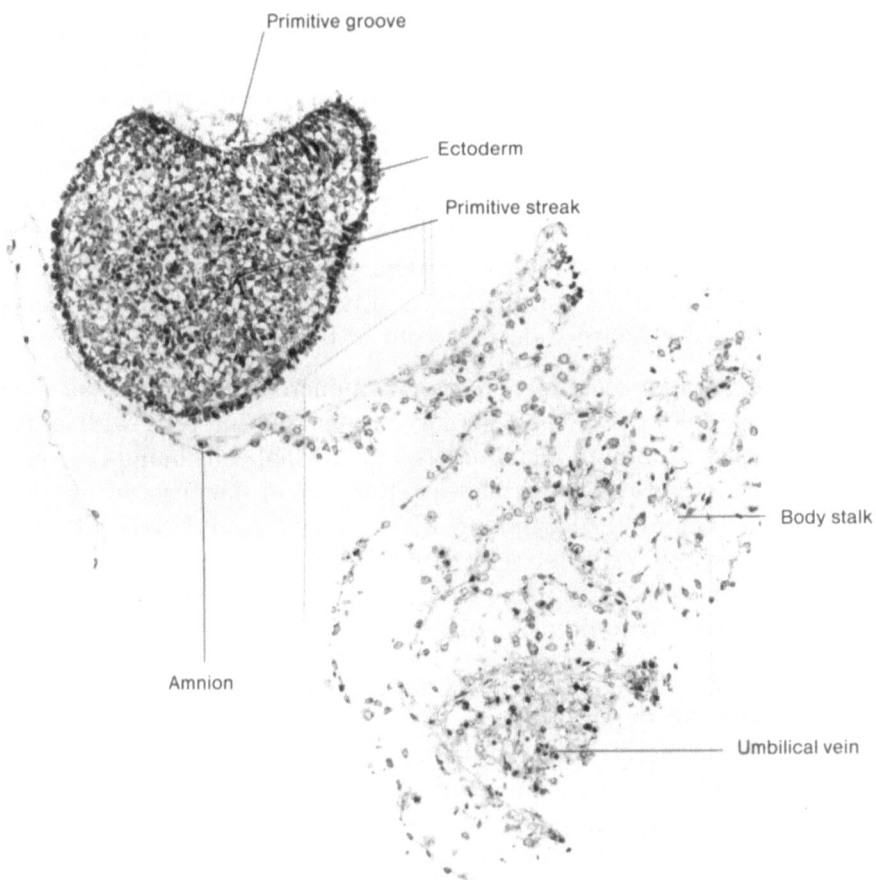

Primitive groove

Ectoderm

Primitive streak

Body stalk

Amnion

Umbilical vein

Fig. 57. Section through caudal end of body (162–3). A shallow depression at the dorsal surface of the body indicates the primitive groove, the lining cells of which lose the basal limiting membrane and intermingle ventrally with densely packed mesenchymal cells of the primitive streak

5. *Marginal layer*. The presence of the marginal layer is widely accepted as the first sign of neuronal differentiation. Müller and O'Rahilly (1986) stated that at stage 11, a cell-free layer is distinct in the lateral walls of RhD and of the spinal cord, but is less developed in RhA to RhC. In our embryo the wall of the neural tube consisted everywhere exclusively of a tall pseudostratified columnar epithelium, i.e., matrix or ependymal layer, and even in the caudal half of the rhombencephalon, where the earliest neuronal differentiation of the whole central nervous system takes place, no sign of differentiation of the marginal layer was perceived. In this respect our embryo showed a retarded development of the central nervous system.

6. *Primordial germ cells*. In our embryo numerous primordial germ cells were found, taking the form of large round lucent cells with a large spherical nucleus among the entodermal epithelial cells lining the ventral half of the hindgut in an area between the level of the posterior intestinal portal and that of the beginning of the allantois. Politzer (1928) and Fujimoto et al. (1977) dealt briefly with these cells in an 18-somite and a 14-somite embryo respectively, but no other authors have referred to them in their description of embryos in stage 11.

7. *Somites*. The 15 pairs of somites recognized in our embryo were classified into four groups according to the differentiation grade and showed a progressive advancement in a craniocaudal direction. A distinct myocoele was seen in the 6th–15th somites but not in the first to fifth somites. In the latter, differentiation into the sclerotome and the dermomyotome has already taken place. This condition was quite similar to that of a 17-somite embryo described by Atwell (1930) but not to that of a 14-somite embryo reported by Heuser (1930), in which each somite contains a myocoele.

8. *Intermediate mesoderm*. In his 14-somite embryo, Heuser (1930) recognized the nephrogenic cords between the levels of the 5th and the 14th somites. They reached their maximum size and differentiation between the levels of the 9th and the 12th somites, consisting of tubules, vesicles, primitive excretory ducts, nephrostomal canals, and one quite definite external glomerulus. In our embryo cell groups of the intermediate mesoderm were encountered between the levels of the 7th and the 15th somites, but indications of pronephric differentiation were detected nowhere throughout the whole length. In this respect, our embryo is less developed than Heuser's 14-somite embryo.

9. *Cardio-vascular system*. The developmental state of the cardio-vascular system in our embryo was very similar to that in Heuser's 14-somite embryo, but the second aortic arch artery, present in his embryo, was not encountered in our case. Comparing the photomicrographs in Streeter's survey (1942) with ours (Figs. 17–20), myocardial differentiation progressed more in our embryo than in a 16-somite embryo (C. C. 7611). Con-

cerning the epicardium, Streeter stated that "the visceral surface of the coelomic passage does not exist as a membrane, separate from the myocardium;" and that "the surface cells do not exist as a distinct and separate layer." This was also true for our embryo, except for a small area of the most cephalic portion of the bulbus cordis, where the simple columnar epithelial cells covered the myocardium as a distinct layer continuous with the pericardiac epithelial lining.

10. Dorsal flexure of the embryonic body. This condition, sometimes strikingly conspicuous, was often found in early-stage embryos: for example, the 15-somite embryo of Dorland (1922), the 17-somite embryo of Atwell (1930), the 18-somite embryo of Politzer (1928), and the 20-somite embryo of Davis (1923). On the other hand, there were also numerous embryos which showed no such dorsal flexure: the 8-somite embryo of Heuser and Corner (1957), the 10-somite embryo of Corner (1929), the 10-somite embryo of Gasser (1975), the 13-somite embryo of O'Rahilly et al. (1982), the 13–14-somite embryo (Pfannenstiel III) of Low (1907–1908), the 14-somite embryo of Heuser (1930), the 14-somite embryo of Nishimura et al. (1974), the 16-somite embryo (C. C. 7611) of Müller and O'Rahilly (1986), and the 17-somite embryo of Wen (1928). Some authors regarded this dorsal flexure as an artifact. Bartelmez and Evans (1926) stated, "as it occurs where the body of the embryo is very thin and not supported by an adjoining massive part, we believe that the collapse of the yolk-sac, which usually caused at fixation, is a more important factor." Our embryo also showed a slight but distinct dorsal flexure centered on the seventh and eighth somites. We found no sound basis for regarding this dorsal flexure as an artifact.

5 Summary

The morphological features of a well-preserved human embryo having 15 pairs of somites are described and illustrated with a complete set of photomicrographs. This embryo was found during a forensic dissection of a Japanese woman, fixed in 10% formalin for 5 days, and embedded in an epoxy resin mixture according to routine procedures. Serial sections about 0.75 μm thick were made and stained with toluidine blue. The most important features were as follows:

1. The embryo measured 4.1 mm in greatest length and was quite symmetrical viewed dorsally.

2. Closure of the neural tube took place from the level of the first branchial groove, corresponding to the level of the anterior one-third of the rhombencephalon, to that of about 0.4 mm posterior to the 15th somite, coming across the anterior end of the cloacal membrane. The neural plate and the wall of the neural tube consisted exclusively of a pseudostratified columnar epithelium, and neither the mantle nor marginal layers were identified anywhere.

3. At the anterior end of the embryo there was a conspicuous optic evagination, and at the dorsal end of the second branchial groove a round otic placode.

4. Of the three segments of the entodermal tract the midgut was the largest, constituting about half the entire length, and opened ventrally into the large yolk sac, whose surface was covered by a meshwork of highly developed blood vessels. In the foregut two branchial pouches were seen, corresponding to the branchial grooves. Primordia of the thyroid gland and of the liver were also recognized. Among the entodermal epithelial cells lining the ventral half of the hindgut anterior to the origin of the allantois were found numerous primordial germ cells.

5. Among the 15 pairs of somites, the first two were small and consisted of dermomyotome of epithelial cell arrangement and of sclerotome mingled into mesenchyme. The next three were large, triangular on transverse section, and consisted of dorsolateral dermomyotome and ventromedial

sclerotome representing a densely packed mesenchymal cell aggregation. The sixth to the tenth somites showed a well-defined triangular contour, each containing a lumen, the myocoele. The last five somites were progressively smaller and were quadrangular in shape.

6. The intermediate mesoderm was encountered at levels from the 7th to the 15th somite, but no indications of pronephric differentiation were detected.

7. The heart tube was relatively large, took an S-shaped tortuous course, and occupied almost the entire pericardiac cavity. It consisted of the bulbus cordis, the ventriculus, and the atrium, each of which consisted of a thick myocardial tube and a thin endocardial tube. The latter was located along the axial portion of the former and separated from the former by a wide transparent space containing no visible coagulum.

Acknowledgements

Cordial thanks should be expressed to Prof. Dr. Y. Mizoi and Dr. T. Fukunaga, Department of Forensic Medicine, Kobe University School of Medicine, who generously afforded the opportunity to investigate this embryo. Thanks are also due to Mr. Y. Kikui for his technical assistance in preparing the photomicrographs.

References

Atwell WJ (1930) A human embryo with seventeen pairs of somites. Contrib Embryol Carnegie Inst 21: 1–24

Bartelmez GW, Evans HM (1926) Development of the human embryo during the period of somite formation, including embryos with two to sixteen pairs of somites. Contrib Embryol Carnegie Inst 17: 1–67

Corner GW (1929) A well preserved human embryo of 10 somites. Contrib Embryol Carnegie Inst 20: 81–102

Davis CL (1923) Description of a human embryo having twenty paired somites. Contrib Embryol Carnegie Inst 15: 1–51

Dorland WAN, Bartelmez GW (1922) Clinical and embryological report of an extremely early tubal pregnancy; together with a study of decidual reaction, intrauterine and ectopic. Am J Obstet Gynecol 4: 215–227, 372–386

Fujimoto T, Miyayama Y, Fuyuta M (1977) The origin, migration and fine morphology of human primordial germ cells. Anat Rec 188: 315–330

Gasser RF (1975) The 10-somite embryo, stage 10. In: Atlas of Human Embryos. Harper and Row, Hagerstown

Heuser CH (1930) A human embryo with 14 pairs of somites. Contrib Embryol Carnegie Inst 22: 135–153

Heuser CH Corner GW (1957) Developmental horizons in human embryos. Description of age group X, 4 to 12 somites. Contrib Embryol Carnegie Inst 36: 29–39

Low A (1908) Description of a human embryo of 13–14 mesodermic somites. J Anat Physiol London 42: 237–251

Müller F, O'Rahilly R (1986) The development of the human brain and closure of the rostral neuropore at stage 11. Anat Embryol 175: 205–222

Müller F, O'Rahilly R (1987) The development of the human brain, the closure of the caudal neuropore, and the beginning of secondary neurulation at stage 12. Anat Embryol 176: 413–430

Nishimura H, Tanimura T, Semba R, Uwabe C (1974) Normal development of early human embryos: Observation of 90 specimens at Carnegie stages 7 to 13. Teratology 10: 1–8

O'Rahilly R, Müller F (1985) The origin of the ectodermal ring in staged human embryos of the first 5 weeks. Acta Anat 122: 145–157

O'Rahilly R, Müller F (1987) Developmental stages in human embryos. Carnegie Institution, Washington

Politzer G (1928) Über einen menschlichen Embryo mit 18 Ursegmentpaaren. Z Anat Entwicklungsgesch 87: 674–727

Streeter GL (1942) Developmental horizons in human embryos. Description of age group XI, 13 to 20 somites, and age group XII, 21 to 29 somites. Contrib Embryol Carnegie Inst 30: 211–245

Wen IC (1928) The anatomy of human embryos with seventeen to twenty-three pairs of somites. J Comp Neurol 45: 301–376

Subject Index

101

Advances in Anatomy, Embryology and Cell Biology

Editors: F. Beck, W. Hild, W. Kriz, R. Ortmann, J. E. Pauly, T. H. Schiebler

Springer-Verlag Berlin
Heidelberg New York London
Paris Tokyo Hong Kong

Springer

Advances in Anatomy, Embryology and Cell Biology

Editors: F. Beck, W. Hild,
W. Kriz, R. Ortmann,
J. E. Pauly, T. H. Schiebler

Springer-Verlag Berlin
Heidelberg New York London
Paris Tokyo Hong Kong